Global Telecommunications Market Access

For a listing of recent titles in the *Artech House Communications Law and Policy Series*, turn to the back of this book.

Global Telecommunications Market Access

Jennifer A. Manner

Artech House
Boston • London
www.artechhouse.com

Library of Congress Cataloging-in-Publication Data
Manner, Jennifer A.
Global telecommunications market access / Jennifer A. Manner.
p. cm.—(Artech House communications law and policy series)
Includes bibliographical references and index.
ISBN 1-58053-306-X (alk. paper)
1. Telecommunication—Marketing. 2. Telecommunication—Law and legislation.
3. Competition, International. I. Title. II. Series.
HE7631 .M363 2002
384—dc21 2002019674

British Library Cataloguing in Publication Data
Manner, Jennifer A.
Global telecommunications market access.—(Artech House communications law and policy series)
1. Telecommunication policy 2. Telecommunication—Law and legislation
I. Title.
384
ISBN 1-58053-306-X

Cover design by Gary Ragaglia

Disclaimer
The views reflected in this book are solely those of the author and do not necessarily reflect the views of WorldCom, Inc.

685 Canton Street
Norwood, MA 02062

International Standard Book Number: 1-58053-306-X
Library of Congress Catalog Card Number: 2002019674

10 9 8 7 6 5 4 3 2 1

This book is dedicated to the most important people in my life. First, to my husband, Eric Glasgow, whom I love dearly and whose constant love and support enable me to pursue all of my dreams. Second, to my parents, Joyce and Maurice Manner, who provide me with the love and encouragement I need to succeed.

Contents

Preface

The competitive landscape of the global telecommunications market has changed dramatically over the past couple of decades. These changes have led the international telecommunications market to evolve from primarily a market consisting of monopoly service providers to one where choice has become a reality for consumers in many markets. This ability to choose among multiple service providers has resulted in tremendous societal benefits, including lower prices for consumers, greater technological innovations, and increased reach of services to more and more consumers. This book focuses on how this progression occurred, where it is going, and how best to get there in a manner that effectively continues to encourage competitive telecommunications markets, with a particular focus on the regulatory and business issues that must be addressed in this endeavor.

Overall, the aim of this book is to provide the reader with a firm understanding of the regulatory, economic, business, public policy, and other considerations associated with entry into global telecommunications markets from both a commercial and government perspective. It also provides an overview of the tools that are available to service providers, governments, and others in ensuring that a competitive telecommunications market is created and fostered on a day-to-day basis.

Acknowledgments

This book could not have been written without the assistance of numerous friends, colleagues, and family members, including my husband, Eric; my parents; and my dear friends and colleagues, among them Donna Bethea, Roya Talazedah, and Diana Erbsen. Each provided me with the support I needed to complete this book.

The idea for this book came after I spent a semester in 2000 teaching International Telecommunications Regulation at Georgetown University Law Center. It was then that I realized there was no current textbook or source that provided the sort of information my students, both foreign and American, would need to know if they were to practice international telecommunications law with regard to market access. Accordingly, the catalyst to this book was developed. To that end, I would like to thank my students; my coprofessor, Jennifer Warren; my research assistants, Stacey Farci and Maura Jeffords; and Dean Martha Hoff, for their inspiration, assistance, and support for this effort. Dean Hoff provided me with encouragement for my book and funding for my research assistants so that this project could be a success.

I would also like to thank my colleagues at WorldCom, Inc., for their support as I drafted this book and the ability to talk through certain ideas both in the abstract and in more concrete terms with many of them. In particular, I would like to thank Mary Sue Shore for her input, especially in regard to investment projects and Africa, and Dan delaMare-Lyon for his technical photography.

In addition, I would like to express my thanks to Professor Rob Frieden for providing me with permission to reprint in my book his glossary of communications terminology.

I would also like to thank my parents, Maurice and Joyce Manner, for reviewing each chapter for me and providing valuable suggestions. I would especially like to thank my mother for her tireless efforts researching all the Web sites of companies and other entities mentioned in this book.

Most of all, I would like to thank my husband, Eric Glasgow. Eric had perhaps the hardest job of all—being that of reviewer, cheerleader, and idea bouncer, all while starting a new position as a Professor of Neurobiology at NEOUCOM. Without Eric's love and support, this book would not have become a reality.

1

Global telecommunications law, policy, and regulation

This book will explore how governments can best achieve a competitive telecommunications market, the impact of the failure to do so, and how new entrants can evaluate a market and structure a successful strategy for market entry. Chapter 1 provides an overview of the critical areas that will be addressed throughout this book, including:

- The components of a competitive market;
- The current status of effective competition in global telecommunications;
- Pressures on the creation of effective competition in telecommunications markets;
- The impact of failing to open a telecommunications market to effective competition;
- The effect of market openness on the creation of a viable telecommunications business and its customers;
- The requirement of individual markets to address these needs in structuring their telecommunications markets.

Subsequent chapters will discuss each of these themes in more detail.

1.1 Introduction

The global telecommunications market is currently one of the most dynamic market sectors. This is in stark contrast to the relatively sedate position of this sector just two short decades ago, when global revenues for the telecommunications sector were less than a small percentage of this number. The majority of changes leading to this dynamism occurred since the mid-1980s, the date when competition was first introduced into the international telecommunications market. This occurred through the institution of new laws, regulations, and policies. Prior to this time, the telecommunications sector was generally characterized much like a typical utility: a single monopoly service provider in each market that was slow to respond to basic customer needs and demands, with standard solutions; the introduction of few innovative technologies and services; and high prices, which funded overly large workforces and other government programs. In other words, most incumbent, monopoly telecommunications service providers were traditionally lethargic, bloated bureaucracies.

Over time, however, the increasing globalization of business, as well as the growing sophistication of computer technology, led to the demands of businesses and other consumers for greater reliability and more reasonable rates for services. In order to meet these and other demands, governments recognized the need for substantial private capital to improve and modernize the telecommunications infrastructure and introduce new services. However, the increase in data traffic also created additional, more complex infrastructure needs. Accordingly, governments began to contemplate both the privatization of their monopoly service providers and the introduction of new entrants into the telecommunications sector.

While at first only a few highly industrialized nations opened their telecommunications sectors to effective competition, these trailblazing markets produced mounting evidence that effective competitive markets provided better, cheaper, and more innovative services to customers, which aided in improving the overall welfare and economy of the country. Eventually, this force fostered the telecommunications revolution that occurred over the past few years. During this period, governments began to recognize that their economies would flourish if they could leverage new telecommunications services and access to improved infrastructure in order to achieve many of their societal goals.

However, this realization did not result in the automatic opening of telecommunications markets. Following these few initial market openers, most governments were still extremely resistant to open the doors to their markets, and even slower to ensure that once the doors were open, effective competition was created. They feared many of the effects of such efforts, including the loss of revenue associated from transferring their government-owned telecommunications systems to the private sector (hence, losing the annual revenues associated with the telecommunications business), the introduction of "foreign" ideas and ways into their markets, a loss of control of a strategic asset, and, in some cases, just the very idea of change.

Fortunately, a few progressive countries that experimented with market opening and crafting regulatory regimes allowing competition to flourish achieved many positive results, including more reliable service, new service offerings, and lower rates, while providing safeguards to mitigate the fear factor inherent in such change. As discussed in Chapter 2, the United States, the United Kingdom, Chile, and Australia all were early market openers. Other countries soon began to follow, but only a few were committed to truly creating a competitive marketplace. Many governments only opened a small market segment, and even then, by failing to constrain the monopoly power of the incumbent, stymied competition in these few open sectors. Still other governments moved forward with privatization of their incumbent service provider, but failed to provide commitment to further market opening. Accordingly, the successes that were seen in the early market-opening countries were not always apparent in these partial openers.

In the past few years, however, this hesitant mindset has begun to change more and more as the positive impact seen from the opening of the global telecommunications market has become apparent. Today, every major developed economic market has introduced some form of competition in the basic telecommunications sector as well as most other sectors. Even many developing countries have at least partially privatized their incumbent service provider and have allowed some limited form of competition. The benefits in almost every market are apparent—lower rates, greater choices, and innovative services, to say the least.

In many countries the impetus for changes in the market that lead to competition are similar, but the direction and impact of the introduction of competition in a country's telecommunications market is directly dependent upon the individual laws, policies, and regulations that govern that market. These differences should not impede the goal of opening a truly effective competitive marketplace, since only then will a government be able to achieve most of its goals, both direct and subsidiary, associated with a vibrant telecommunications services market. As will be explored later, many governments refuse to open their markets or only pay lip service to the concept of competition. Governments often mistakenly believe that their national security, cultural identity, and financial well-being will be compromised by further market opening.

The contrary is actually true; as more and more nations are opening, governments are finding that many of their early fears were ill-founded, because they can be addressed through the structural safeguards (legal, regulatory, and otherwise) that they institute. Further, markets that are refusing to open are finding that their economy is suffering, since businesses choose to locate in markets with better, cheaper, and more innovative access to the information superhighway.

1.1.1 The current status of the electronic communications sector

Before delving into the process to achieve competition, it is worthwhile to understand briefly the current status of the electronic communications sector and why it is so important to today's economy. One fact remains clear: The telecommunications industry is one the most dynamic markets in history. This has been driven by:

- Technology drivers, including the introduction of computer chips, fiber-optic networks, and new mobile service technology;
- The increasing importance of the Internet;
- The growing competitive environment and global nature of the economy;
- The new paradigm for companies in today's international and high-technology business world;
- Changing customer demands and applications—the need for data service anytime, anywhere;
- Fluid market dynamics, including the "dot.com" craze of the late 1990s and the subsequent "dot.com meltdown."

These drivers have led companies to become more dependent on data services and the ability to have seamless communications wherever the locations of their offices; they have changed the traditional trade paradigm, as seen in Table 1.1. For example, Telecordia Technologies has estimated that a significant and growing percentage of all traffic carried by telecommunications networks will be data [1]. Further, there has been a

TABLE 1.1 Trade Paradigms

OLD TRADE PARADIGM	NEW TRADE PARADIGM
Services	Agriculture and commodities
Industrial goods	Services
Agriculture and commodities	Software
Manufacturing	Banking
Oil and gas	Communications infrastructure
Military goods	Industrial goods
	Transport
	Insurance
	Manufacturing
	Oil and gas
	Military goods
	Aerospace

dramatic growth in demand for international telecommunications services over the past few years, as demonstrated by the recent proliferation of international service providers and networks.

In order to meet this new paradigm, telecommunications service providers must locate their businesses where they can be the most responsive

TABLE 1.2 Examples of Geographic Reach

STABLE BUSINESS ENVIRONMENTS	REASONABLE LABOR COSTS	AVAILABILITY OF SCARCE RESOURCES
United Kingdom	India	Venezuela
France	China	Kazakstan
Japan	Taiwan	Trinidad and Tobago
Germany	Indonesia	United States
United States		Saudi Arabia
Finland		Russia
		Iran
		Saudi Arabia
		Kuwait
		United Arab Emirates

to their customers' needs. Accordingly, the geographic reach of service providers grows as businesses become more global in nature. Traditionally, as outlined in Table 1.2, companies have sought to locate their offices in countries where there have been stable business environments, where scarce resources are located that are critical to their operations, or where labor is cheap or highly skilled. However, as more and more markets are open to the sale of goods and services, the reach of these businesses has continued to expand to more untraditional markets, such as Vietnam and Israel.

These considerations and this background should be kept in mind as this book progresses in order to understand why certain market access decisions are made by service providers.

1.2 The process of achieving competition: privatization, liberalization, and effective regulation

1.2.1 The basic formula to achieve effective competition

In order to examine the global telecommunications market, it is essential to understand the formula necessary to achieve a competitive telecommunications market. As discussed in detail in Chapter 2, the formula is compromised of three basic elements:

Privatization + Liberalization + Effective Regulation = Competition

Each of these elements must be reflected in the law, policies, and regulations of a country in order for effective competition to be created in that telecommunications market. However, mere words on paper, as we will see in later chapters, are insufficient to ensure the creation of competition. Accordingly, in order to achieve the goal of true competition, the government must be committed to act to ensure that each element in the formula for a competitive telecommunications market is fully implemented in a meaningful manner.

Although this formula and its implementation will be the subject of greater discussion in future chapters, it is helpful to understand the meaning of each element of the competition basic formula. Each term will be utilized and relied on extensively throughout this book, since they are the cornerstones to understanding, structuring, and entering the global telecommunications market.

The definition of privatization

Privatization in this context refers to a country eliminating complete or partial government ownership in its telecommunications service provider(s). As discussed in Chapters 2 and 3, originally almost all countries had government-owned telecommunications service providers that were granted a monopoly over the provision of all telecommunications services within the country as well as their international communications. However, as new services (e.g., computer services, including e-mail and the Internet) were introduced, and the world grew smaller in part because of improved telecommunications services, telecommunications became more important to the economy and welfare of each country's citizens, with a new demand for even more services and greater access. In order to satisfy these needs, improved infrastructure and new technologies were required. However, this often required an infusion of large amounts of capital and the introduction of new technologies and services. In the 1980s, governments began to recognize that a private entity might best be able to manage the new requirements of modern telecommunications, because of its greater expertise in management of a commercial enterprise and its access to private capital.

This led, initially, to a few trend-setting countries, such as the United Kingdom and Chile, privatizing their monopoly government service providers. In these markets, the change was noticeable. For example, in Chile, competition and privatization encouraged the incumbent service providers to increase their national coverage and telephone density, as well as improve their services and lower their rates. After this success (and that of other countries) was widely noted, there began a movement to privatize, at least in part, the incumbent telecommunications service providers in several key markets, including Germany, France, South Africa, and Ghana. Many of these were highly successful and continue to serve as a model for privatization for other countries. Some other key markets, however, such as Japan, have only gone through a partial privatization. And other important markets, such as the United Arab Emirates and Thailand, as discussed later, have refused to date to establish a timetable to privatize their incumbent service providers. In such markets, the telecommunications and other associated sectors have suffered, since many of the traditional hallmarks of a monopoly environment have continued to reign.

The definition of liberalization

Liberalization is the process of allowing competitive service providers into a specific sector of the telecommunications industry or into the telecommunications industry as a whole that has formerly been closed to new entrants. As discussed, most markets originally were structured as a monopoly for telecommunications services (or in certain rare cases, duopolies). Similar to privatization, the concept of multiple providers of the same telecommunications services did not really begin to blossom until the late 1980s. The concept of liberalization really only became a reality in the telecommunications arena when the long-distance market in the United States became competitive and when the United Kingdom began to open its telecommunications market to competition. In these countries, multiple service providers were allowed to enter into previously closed market sectors, most notably for value-added services and long-distance service. The results of this market opening were readily apparent: lower rates, greater choices, innovative services, and better customer service were achieved, while the fears typically associated with market opening did not materialize.

This encouraged some increased market opening but at first only incrementally. The late 1980s saw many countries that were hesitant to liberalize more than the value-added service sector. Accordingly, until recently, many markets that had partially liberalized were characterized as having one basic service provider, possibly privatized, but multiple providers of value-added services, such as facsimile or data services.

Liberalization of the basic services sector only developed on a wide-scale basis once success was demonstrated in a few additional key markets that undertook this endeavor. For example, many of the Latin American markets were hesitant to move forward with the liberalization process until some of the larger and more economically advanced large regional markets of Brazil, Chile, and Argentina succeeded in taking this step. Unfortunately, to date little market opening has really taken hold in the lesser-developed markets, such as the countries that comprise Western Sahara Africa. It is likely that full global liberalization will only occur once a few leading nations among the lesser-developed countries truly open their markets to effective competition, and the positive results are well documented and disseminated.

The definition of effective regulation

Effective regulation is the cornerstone of achieving a competitive telecommunications market. It is not, however, regulation for regulation's sake. Essentially, effective regulation is the creation of a rational regulatory scheme, which provides for a transparent process and allows the regulator to enforce its rules in an effective manner. Key concepts that should generally be included in structuring a competitive telecommunications market include:

- Creating an independent telecommunications regulator;
- Mandating transparency in the regulatory and legal processes;
- Imposing dominant service provider safeguards;
- Requiring cost-based interconnection;
- Imposing nondiscrimination requirements;
- Ensuring that an appeals process is available from decisions of the regulator to a neutral body;
- Establishing clear authorization requirements, including identifying the type of process to obtain an authorization (e.g., auction, beauty contest, straight rules, registration, comparative hearing) and any terms and conditions of the authorization;
- Eliminating or minimizing foreign investment and operational control limits;
- Stating clearly any local presence requirements and ensuring that these requirements are easy to meet;
- Establishing clear and strong regulatory and judicial enforcement mechanisms;
- Imposing a fair tariff regime and process;
- Providing access to scarce resources, including spectrum, access to buildings, easements, and rights of way;
- Creating a neutral funding mechanism to ensure universal service;
- Imposing unbundling requirements.

It is important that these concepts be brought into the regulatory regime in a manner that will encourage the creation of competition but is balanced with the specific requirements or goals of the overall country. Achieving this approach will be the subject of many later portions of this book.

1.3 Current status of effective competition

To date, almost all developed countries and a few developing countries have begun the process of opening their telecommunications markets to effective competition. Some countries, such as the majority of European Union (EU) member states and the United States, have made far-reaching strides in creating a truly competitive telecommunications market for almost all services. Other countries, such as those in the Middle East, including the important markets of Saudi Arabia, Iran, and the United Arab Emirates; most African nations; and some key Asian markets, such as Vietnam, have not even begun the process of opening their markets to any form of competition nor do they have any firm plans to do so.

An important indicator of the level of market opening that is occurring around the globe is the number of national telecommunications authorities that have been created in the past decade. Generally, in closed markets, the government and the service provider are the same entity. Accordingly, there is no separate regulatory body to oversee the telecommunications market. In the 1990s, for example, the number of telecommunications regulatory bodies increased substantially [2]. This dramatic jump demonstrates the increasingly popular form of privatization of incumbent telecommunications providers but only hints at the opening of the market and the imposition of effective regulatory structures around the globe.

Overall, as of late 1999, about 73% of global basic telecommunications markets were still closed to competition [3]. It is likely, however, that over the next decade this percentage will drop dramatically, since many countries have already publicly stated their intention to open initially or further to competition. For example, the Brazilian government plans to expand its duopoly for basic long-distance and local services to full competition in 2002. Similarly, the South African government plans to open the basic service market to competition in 2003, when the Telkom South Africa monopoly ends. Despite this planned growth, it is likely that many lucrative markets will still remain effectively closed for years to come, most notably many of the Middle Eastern states and many of the developing countries in Africa, Eastern Europe, and Asia. For example, key markets, such as the United Arab Emirates, have not made any clear announcements about plans to open their markets to competition.

It is important to remember that basic services are the last to open up to competition. Governments generally are most willing to open new telecommunications service markets to competition when this will not impact the predicted and expected revenue base of the incumbent service provider, which is often government owned and recognizes such revenue. Accordingly, the first markets that today are generally opened to some form of competition are the mobile service and Internet service markets. For example, the International Telecommunication Union (ITU) has found that 67% of telecommunications markets had some form of competition in the mobile services, and 72% had some form of competition in their provision of Internet services [4].

In regards to mobile services, most governments have allowed some form of competition into this market. However, what this does not reflect is whether or not a market has instituted an effective regulatory regime governing the provision of this service. For example, competition concerns have been raised primarily in the interconnection of mobile service providers with fixed service providers. Generally, the basic service providers, which control the bulk of the domestic telecommunications network, have used this power to try to limit the success and reach of the burgeoning mobile services market [4]. This type of complaint is not limited to markets in the beginning stages of competition; on the contrary, this behavior has been well documented in the EU and other liberalized markets.

In terms of Internet services, although this market has generally been open to some form of competition, there have been tremendous efforts in many countries by incumbent service providers to limit the ability of Internet service providers to provide service in an economical and far-reaching manner. For example, in South Africa, Telkom (the incumbent basic service provider) sued the regulator for issuing Internet service authorizations, arguing that it was done in violation of its basic service exclusivity [5]. To date, the courts and the regulator have sided with the Internet providers, but the case is still pending on appeal before the highest court in South Africa. Despite the judicial success of the Internet service providers, Telkom has consistently tried to limit these providers' ability to obtain capacity for the provision of their services.

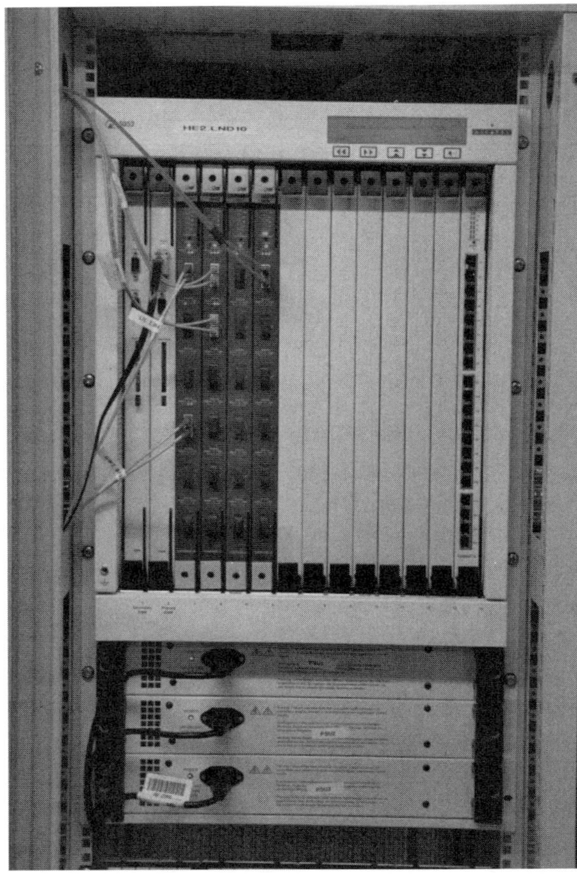

Figure 1.1 Alcatel packet engine: driving data across the globe.

Another key market where competition has been able to blossom is the value-added service (VAS) market, including services for high-speed data, such as Frame Relay, as well as less-advanced services, such as facsimile and voice mail. Figure 1.1 illustrates a typical piece of data network equipment. In terms of these services, many governments have moved forward fully with the liberalization process, because they have found it an efficient way to bring advanced services into their market—to satisfy the needs of business users, while still preserving the revenue base obtained from the basic service provider. Since the basic service markets in many countries remain closed, the VAS providers in these markets must obtain all capacity (local, national, and international) from the incumbent basic service providers. This provides the basic service providers with a large degree of leverage and control over the VAS market. In the more lucrative service markets, such as high-speed data services (e.g., FRS and ATM), these types of service providers have been seen as a threat to the basic service providers. Accordingly, there have been many incidents of incumbent service providers cutting the circuits of VAS providers, because, they allege, the provision of these services violates their exclusivities. One notable case is the past actions of Telecom Colombia, which, in the 1980s, consistently alleged that the VAS providers that provided FRS were violating its monopoly over basic services.

1.3.1 Pressures on the creation of effective competition

As will be discussed throughout this book, there are many outside influences that drive the creation of competitive telecommunications markets. First and foremost are the economic realities associated with open telecommunications markets—overall increased well-being for the country's economy. Further, many markets are recognizing the benefits of their own population having access to greater telecommunications services as a means to deliver tele-education, tele-medicine, and other services. Other countries recognize the importance of building a more developed, modern, and far-reaching infrastructure. Nonetheless, many governments, while recognizing these benefits, still fear increasing private ownership, because of the loss of control of strategic assets and access and the similar effects that are brought about by effective competition. As discussed throughout this book, many businesses (both telecommunications service providers and their business users) spend a substantial amount of resources to try to explain the virtues of competition and the procedures available to these governments to ensure that their concerns are addressed in the regulatory scheme.

However, additional outside pressure (generally political in nature) is often required to jump-start market opening and encourage the development of effective competition. As will be discussed further, regional or international pressure is often a driving force for this change to occur. In this regard, both international and regional organizations play an important part. The most far-reaching was the 1998 World Trade Organization (WTO) agreement on basic telecommunications services [6]. This multilateral agreement (and the negotiations leading up to it) made great strides in obtaining commitments from many governments around the world to open their telecommunications markets to competition in varying degrees. The initial negotiation process provided a great deal of pressure on countries to formalize and commit to liberalization milestones. Further, the commitments agreed to in the WTO may provide leverage for countries to force market opening and create effective regulatory structures throughout the liberalization process. There is a follow-on agreement that is being negotiated that will result in even further market opening.

There are also other international and regional organizations, which are discussed throughout this book, that are very active in encouraging the creation of competitive telecommunications markets. These include efforts by the ITU, the Organization for Economic Cooperation and Development (OECD), InterAmerican Telecommunications Commission (CITEL), and the Asia Pacific Telecommunications (APT) organization, among others. These organizations will often issue policy documents on the need for competition in the telecommunications sector and methods for achieving it, as well as structuring programs to assist governments in achieving the privatization, liberalization, and regulatory goals.

In addition, smaller and less well-known mechanisms exist. For example, many free trade agreements have helped to advance liberalization, often on a regional basis. The European Free Trade Area (EFTA) has helped to open the non-EU countries of Western Europe to full competition and the creation of effective regulatory regimes. The EFTA nations believed that they would miss out on the telecommunications revolution to their more open neighbors. Today, every nation in Western Europe, either by virtue of its EU membership or its EFTA affiliation, has opened its market to competition. Other groups, such as the United States Telecommunications Training Institute (USTTI), which is partially privately funded and partially funded through the U.S. Department of States AID Program, provide training sessions for regulators on all aspects of regulation. These training programs often provide a sound basis for regulators from developing telecommunications markets to understand how to structure their own telecommunications market in a manner that is procompetitive yet recognizes their cultural identity.

The World Bank also has played an important role by financing the efforts of many nations seeking to improve their economic situation through telecommunications. Specifically, the World Bank's global information and communications section provides policy advice, investment support for the public subsidy of privately owned provider infrastructure to rural households and the urban poor, and investment support for private providers of fixed and mobile service providers and other similar services on a global basis to countries in need.

Finally, another important pressure point is that of individual government contact. This may take the form of direct bilateral negotiations on telecommunications-related issues. Governments, in the context of other discussions, may also discuss telecommunications issues. For example, in exchange for certain concessions on a government's part in telecommunications, another government may be willing to give up pressuring the first on another key issue.

1.4 The impact of failure to open the market to effective competition

Despite the great strides that have been made in introducing competition into the telecommunications sector around the world, most markets are still closed to any form of competition, and another equally large amount still only pay lip service to the concept of true competition. Many markets, for example, may enact regulations that provide for some form of competition, yet they lack the will or tools necessary to enforce these regulations to ensure that effective competition is created. These less than effective regulations are often caused by a lack of commitment on the part of the government to truly open its market to competition and allow it to prosper.

Ultimately, governments that continue to reject opening their markets to effective competition, as the decade wears on, are likely to see an even greater direct impact on their economy and the well-being of their citizens. As will be discussed, failure to open a telecommunications market to effective competition has very real direct and indirect effects.

1.4.1 Direct impacts of the failure to create competition

In terms of the direct impacts on markets that fail to create a competitive market, as will be discussed further in Chapter 3, these generally include excessively high prices for services and products that are controlled by the monopolist, limited services to users, antiquated services, and poor quality of service. For example, in Rwanda, where the telecommunications market is primarily closed, a population of over 6.5 million has less telephones than most large global corporations. This means that all but a select few have access to any form of telecommunications. Such problems may exist even in a market with limited competition, such as a duopoly, or in an environment where there is no effective form of regulation. Further, the failure to open a market continues the likelihood that large sectors of the population of that country do not have access to modern telecommunications services or, quite possibly, to any telecommunications services.

A brief survey reveals that these problems plague almost all closed markets. The question arises whether, if a lesser-developed market opens to effective competition, it can recognize many of these direct benefits. Lesser-developed countries will often point to examples where limited or partial market opening has not brought the results that the government initially touted. For example, in Zimbabwe, the government created a duopoly for mobile services in Harare and other regions. Nonetheless, poor coverage and high rates still characterize this market. An examination reveals that the process that was utilized was less than transparent, and political favoritism may have played an important part in the authorization process. It is likely, however, that a more fair and transparent process would have resulted in an improved mobile services market in Zimbabwe.

1.4.2 Indirect impacts of the failure to liberalize

In terms of indirect impact, these are likely to be more far reaching. For example, the failure to open a market to effective competition may discourage businesses to locate in a particular country. Large corporations often have requirements for a certain amount of reliability in their networks or the availability of certain sophisticated services, such as Frame Relay Service (FRS), a form of high-speed data. In many closed markets, such as South Africa and Thailand, these new services may not be available or there may be problems associated with these services, such as the failure to have diverse network routing or the inability to provide the required reliability of the network, that limit their attractiveness to consumers. These problems may discourage certain companies from entering a market at all or force them to curtail the way they plan to enter into a market.

Another indirect impact is that the high prices associated with telecommunications markets in a closed market may also dissuade entry by businesses. Often, especially for businesses that are dependent on telecommunications services, such as call centers, these businesses will move to areas where telecommunications services are priced at a much lower level. A good example of this involves Ireland. Ireland originally requested and received a derogation from the timetable of the European Union's requirement to open member state telecommunications markets to full competition. Ireland was partially fearful of the loss of revenue its government would face from having competition. However, shortly after this decision was made, the Irish government began to recognize that it was being hurt more by having a closed market than if it opened its market to competition. For example, Ireland was losing out on obtaining a valuable new business line—that of call centers. Large companies seeking to locate their call centers in English-speaking countries were bypassing Ireland because its telephone rates were too high. Instead, these companies were looking toward England and other English-speaking countries where competition had already been introduced and rates had fallen dramatically. The Irish government, upon recognizing its mistakes, acted quickly and fully opened its market to competition 13 months in advance of its required date. Ireland now has a blossoming call center business and has seen the benefits of competition in its telecommunications markets in many of its other sectors [7].

Another indirect impact is the ability of the country to develop e-commerce. Key components of the e-commerce marketplace are the underlying telecommunications infrastructure and the Internet infrastructure. E-commerce is quite beneficial to an economy because it:

- Allows more information to circulate on available goods and services;
- Introduces high technology into developing markets;
- Leads to lower prices in the marketplace;
- Attracts other businesses that serve e-companies;
- Provides more choices of suppliers and services;
- Provides constant access to goods and services;
- Lays the framework for a low-cost form of entry to market;
- Creates job opportunities for people where the e-company operates.

E-commerce revenue is anticipated to increase around the world. For example, in 2003, the *E Marketer* in July 2000 estimated the global total will be $443 billion.

Many other indirect impacts abound, including the advantages that improved telecommunications services bring to society in general, such as improved access to medicine and education. These will be discussed throughout the upcoming chapters in more detail.

1.5 Overall impact of market openness on the creation of a viable telecommunications business and its customers

The overall impact of a failure to open a market to effective competition affects two general parts of industry: telecommunications service providers and nontelecommunications businesses (e.g., business customers, including businesses such as call centers and dot.coms, both of which are heavily reliant on telecommunications services).

1.5.1 Impact on a viable telecommunications business

Many telecommunications service providers today are hesitant to enter into markets where there is limited or no effective competition. In fact, the mantra of many companies, including AT&T, WorldCom, and Sprint, is that they will only enter into markets where a level playing field has been created and generally where a market has no arbitrary limits on entry. This means that no market player is able to have a regulatory advantage over another player.

These service providers have learned from their own past experiences as well as the past experiences of others that the creation of a level playing field is critical to their ability to succeed in a market. For example, AT&T and WorldCom, Inc., made large investments in the 1990s in the Mexican long-distance telephony market. However, despite certain pledges by the Mexican government to ensure a level playing field in the marketplace, both AT&T and WorldCom's Mexican ventures have been stymied by the anticompetitive tactics of the incumbent service provider, Telmex, and the failure of the Mexican government to take action against Telmex. The numerous anticompetitive problems in the Mexican telecommunications market can be grouped in the following four main categories [8].

1. *Price squeeze tactics.* Telmex engaged in "price squeeze" tactics against its competitors. As the former monopoly service provider, Telmex controls many of the inputs necessary for competing service providers to provide services in Mexico. Telmex charged competing service providers more for these inputs on a wholesale basis than it charged some of its retail customers for carrying the entire call.

2. *Exploitation of bottleneck facilities.* Telmex exploited the bottleneck facilities it controls. Telmex has consistently failed to provide leased circuits to new entrants in a timely manner. Telmex committed to specific delivery times in an agreement signed with other service providers, but it complied with those delivery times in only the rarest of cases. In fact, for a lengthy period of time Telmex refused to provide circuits to competing service providers altogether. Without these circuits, competing service providers are unable to serve customers at all and committed customers will often flee to the incumbent service provider where they are certain they can obtain service.
3. *Failure of the regulator to act.* The Mexican regulator COFETEL has failed to prevent Telmex from engaging in anticompetitive behavior. New entrants have filed hundreds of complaints on various issues with COFETEL, yet it has failed to take action to prevent Telmex from abusing its dominant position. Only since late 2000, when the new Fox administration came into power, has COFETEL been more aggressive at acting against Telmex's abuses. A level playing field requires an independent regulator with the ability and will to enforce the rules.
4. *Failure to liberalize certain key markets.* The Mexican government has allowed Telmex to maintain a monopoly over negotiation of rates for inbound international telephone calls by allowing Telmex a large amount of authority over the negotiation of rates for the termination of inbound international calls. As a result, the cost to terminate calls originating in the United States in Mexico is far above competitive rates. The major reason often ascribed for these high rates is the lack of competitive alternatives for terminating traffic in Mexico.

However, in early 2001, it appeared that these abuses have begun to be addressed; these past actions have had a direct impact on the Mexican economy. For example, customers are being denied choice in the Mexican market for service providers. It is likely that if Mexico were not the largest trading partner with the United States, even more direct and larger impacts would be felt.

Because of past negative experiences, such as in Mexico, many global telecommunications service providers are hesitant to enter into markets where there is not a clear commitment by the government to creating a truly competitive telecommunications market. In this regard, as discussed later, these service providers often seek out markets where a government is far along in the process, is open to suggestions, or will make commitments to the provider in exchange for its entry.

1.5.2 Business customer woes

As discussed throughout this book, in evaluating where to locate their sites, many businesses evaluate the various costs of entering a market before proceeding. Such costs include day-to-day operating costs, including those of telecommunications services. Accordingly, when a company has flexibility in where to locate its business, if a telecommunications service will drive up costs exponentially, they may reevaluate entering that market, in exchange for another, cheaper market. Another important consideration is the level of quality of services available. Many companies, such as banks, require very high service levels, since the transfer of information between global locations is so vital to their businesses. Such a business customer may be hesitant to enter a market where reliability of communications is severely lacking, which is often the case with closed markets.

Similarly, business customers are often hesitant to expand business where they have had a negative experience. Customers may have no choice but to operate their business in a country with limited access to telecommunications services because of strategic importance to the business. However, future expansions may be placed at jeopardy without the required telecommunications infrastructure to support such movement. Accordingly, the telecommunications market is often one of several key considerations, especially for telecommunications-reliant businesses, in where to locate a site. As discussed subsequently, the decisions of customers directly impact whether a service provider will seek to enter a market.

1.6 Individual markets must address their specific needs in creating a competitive market

Ultimately, while the basic formula for creation of effective competition in a telecommunications market must be met for success, there are many ways in which to utilize this formula to reach the desired outcome of an individual government. It is imperative for the formula to be successful—that each government creates a legal, regulatory, and policy structure that fits in with that country's individual needs, including its cultural requirements and legal regime. This means that in some countries a more direct path to competition is possible, while other countries may require a short period of limited competition before opening its market fully to competition. In still others, a full market opening or an accelerated opening for a very short period of time may be most appropriate. However, too long a transition period will have negative results, as well imposing a plan that is not flexible or not able to change to address new circumstances. For example, in Brazil, the government believed that in order to open its basic telecommunications market, it was best to privatize the market and allow limited competition simultaneously. Accordingly, in the late 1990s, the government initiated a process whereby it privatized its monopoly service providers for long-distance and local service and almost simultaneously auctioned off licenses for mirror service providers. These companies were to have a limited exclusive period before full competition would be permitted. By following this approach, the government felt it would be able to manage initial competition while still bringing many of the benefits of competition to its citizens.

A key to remember is that in many countries the legal regime requires more than just regulatory changes in order to open the telecommunications market to competition. Often, whole legal structures must be revised and may include dramatic changes, such as constitutional amendments, enacting laws and creating new organizations, and major restructuring of political entities, such as separating out the regulator from the service provider. This often takes time. As will be discussed throughout this book, the individual nature of each market must be carefully examined before a strategy is chosen, either for structuring the market or for market entry. Accordingly, although some generalization is possible, it's imperative that each market be analyzed in terms of its own laws, regulations, and policies, both for creating a regime to govern the telecommunications market and for determining a market entry strategy. This book will, however, endeavor to provide a general blueprint for most markets on how to structure a sound regulatory, policy, and legal structure for telecommunications market entry.

References

[1] http://www.telcordia.com/newsroom/briefs.

[2] ITU, *Trends in Telecommunications Reform*, 3rd ed., 2000, p. 13.

[3] ITU, *Trends in Telecommunications Reform*, 3rd ed., 2000, pp. 1–7.

[4] ITU, *Trends in Telecommunications Reform*, 3rd ed., 2000, Chapter 1.

[5] http://www.balancingact-africa.com/news/back/balancing-act9.html.

[6] http://www.wto.org/english/tratop_e/servte_e/gbtoff.htm.

[7] 222.odtr.ie/docs/pres060901.doc.

[8] http://www.ustr.gov/releases/2000/04/00-25.pdf.

Selected bibliography

Commission Directive 90/388/EEC, "Competition in the Markets for Telecommunications Services," *Official Journal,* L 192, 24/07/1990, June 28, 1990, pp. 0010–0016, http://www.europa.eu.int.

Frieden, Rob, *Managing Internet-Driven Change in International Telecommunications*, Norwood, MA: Artech House, 2001.

Gartner Group, *Liberalization Status Among Telecommunications Markets Worldwide*, Washington, D.C.: Gartner Group, 2000.

Intven, Hank, *Telecommunications Regulation Handbook*, Washington, D.C.: Infodev, 2000.

2

The basic regulatory equation: the foundation for competition

This chapter focuses on the importance of the creation of a competitive telecommunications marketplace through the regulatory process. It specifically examines the basic regulatory equation for achieving competition in the global telecommunications market and why competition is such an important goal for all governments in structuring a regulatory regime for their telecommunications market. This chapter further builds on earlier discussions of the concepts of privatization, liberalization, effective regulation, and deregulation. The conclusion examines three case studies in different world regimes where the goal of a competitive telecommunications market is being achieved and through which governments are able to satisfy many of the subsidiary goals that make a competitive telecommunications sector attractive.

2.1 The basic regulatory equation

As discussed, there is a basic regulatory equation that has led and will continue to lead to the successful creation of a competitive marketplace:

$$\text{Privatization} + \text{Liberalization} + \text{Effective Regulation} = \text{Competition}$$

There is also a possible fourth component of the equation that was not previously discussed: deregulation. However, deregulation is rarely a factor in this formula until the specific telecommunications market has matured to a very advanced state of competition and even then, deregulation generally only occurs in limited sectors of the market.[1] This concept is briefly addressed later in this chapter as well as throughout this book.

As discussed below, each of the basic regulatory equation's components must be achieved if a truly competitive telecommunications market is to be created and sustained. Concrete examples of markets not able to reap the many subsidiary benefits of competition abound, because of a government's failure to implement a specific component either fully or in a meaningful manner.[2] Accordingly, allowing partial competition in a sector of the telecommunications market or in the market itself, while preferable to a market devoid of competition, is still not an effective means of achieving the benefits that competition can bring into an economy or society. Therefore, each component of the basic regulatory equation must be met in order to achieve the benefits of a competitive telecommunications market.

2.2 Why competition?

2.2.1 A brief chronology

Initially, as discussed further in Chapter 3, competition was not the goal of most governments in developing and fostering their telecommunications markets. On the contrary, as with most essential services, including railroads, gas, and electricity, governments believed early on that in order to provide reliable telecommunications services to the greatest number of their citizens, it was essential that telecommunications service providers

1. Many countries will only deregulate a telecommunications sector where there is little potential for abuse by the service provider. For example, one common area that is deregulated receives only very small aperture earth stations. This service requires minimum regulatory scrutiny because the receive-only nature of the communications service eliminates concerns of interference with other radio-based services and limits potential for abuse by users.
2. For example, in late 2000, the government of Poland planned to auction competitive mobile telephony licenses. However, in part because of the lack of an effective regulatory regime, only three applicants submitted notices of intent to bid for the five concessions. In response, the government decided to call off the auctions and offer the three concessions through a straight paper process.

be awarded essentially monopoly status for their entire country market.[3] Similarly, many governments did not want to entrust the provision of an essential service, such as telecommunications, to private operators.[4] Accordingly, most governments established government-owned and government-operated monopoly service providers in the formative years of the telecommunications sector. It would take decades, at least until the early 1980s, for the concept of a private entity providing basic facilities-based telecommunications service, or more than one entity providing these services, to become a possibility in the minds of even the more progressive and reform-minded government leaders.[5] As discussed below, the United States was the leader in this movement, with other markets, such as the United Kingdom and Chile, following closely behind.

There are several reasons that the United States was the first market to allow competition in its telecommunications market. First, its own virtual

3. The United States also took this view in the early stages of the development of its telecommunications market. In the early 1900s, at the start of the telecommunications revolution, there were a multitude of telecommunications service providers operating in the United States. However, AT&T was able to gain control of the marketplace through its holding of essential patents on voice transmissions and later patents on long-distance communications technologies. By leveraging these advantages, AT&T was able to provide superior long-distance service to its affiliates and was able to essentially force independent operators to shut down. The United States government would not challenge this de facto monopoly for many years, although it did recognize the potential for anticompetitive behavior by such a dominant service provider. Partly in response to this potential, the Communications Act of 1934 was enacted. The Act provides that its purpose is "[to] make available, so far as possible, to all people of the United States, a rapid, efficient, nationwide and worldwide wire and radio communications service with adequate facilities and reasonable charges." 47 U.S.C. § 151.
4. Because of the critical nature of telecommunications services to the well-being of a nation, many governments remain hesitant to entrust these services to anyone but a government-owned and government-operated entity. For example, in many countries the incumbent government-owned monopoly service provider is responsible for providing both civilian and military communications applications. Accordingly, in these countries, it may be expensive (e.g., to build a redundant network) and politically difficult to create a separation of the telecommunications service provider function solely for critical government uses, or the government may be hesitant to allow a private entity to control these communications. Other countries, such as some Middle East nations, may view government ownership and control of the communications network as necessary because of the government goal of controlling its populations' communications through monitoring and other means.
5. The opening of value-added services, such as facsimile and voice mail, began slightly before this time frame.

monopoly, AT&T, was privately owned and operated. The fact that there was already a separation between the service provider and the government in the United States meant that the U.S. government did not have a direct financial interest in the service provider that faces most other governments. This allowed the United States to by-pass the very difficult hurdle of privatizing its incumbent service provider in creating a competitive telecommunications market. Second, and of perhaps greater importance, the United States had established an independent and effective regulator in 1934 when it created the Federal Communications Commission (FCC)[6]. Although, as discussed below, the FCC did not always follow a procompetition bent in its rulings,[7] its political independence and its separation from the service provider allowed it to grapple with these issues in a generally fair manner and adapt to court decisions that mandated market opening.[8] Third, and of equal importance, is the concept of entrepreneurship, which is firmly rooted in the American culture. Accordingly, because American entrepreneurs saw how lucrative the telecommunications industry was, they continually made costly legal and business challenges to the AT&T monopoly, even when such challenges appeared futile, in an effort to gain a piece of the large economic pie that AT&T had virtually controlled.

The first major challenge to the monopoly model of telecommunications services provision in the United States was by a small company called Microwave Communications, Inc. (MCI).[9] Beginning in 1959, the FCC

6. The U.S. Congress created the FCC in 1934 through the Communications Act of 1934. 47 U.S.C. § 154 et seq. (1934).
7. In fact, AT&T created a strategy under Theodore Veil of "One Policy, One System, Universal Service" (basing this strategy on the control of technology giving AT&T certain advantages), which helped improve AT&T's relationship with the FCC. Specifically, the universal service aspect of this policy gave AT&T a special relationship with regulators that led to a regulatory compact in the early 1900s, which helped AT&T and its Bell operating companies to ultimately become the dominant service provider of basic telecommunications services in the United States by being willing to meet the increased service coverage requirements sought by the U.S. government.
8. The FCC's primary mandate is to serve the public interest, convenience, and necessity. 47 U.S.C. § 154 (1934).
9. There were several less-obvious attempts even earlier. One of the most notable of these was the Hush-A-Phone case. In this case, AT&T alleged that a company that had manufactured and sold a simple attachment to the customer handset had violated its monopoly on the telecommunications network. The FCC agreed with AT&T. Ultimately, the courts heard this case and the FCC's decision was reversed. *Hush-A-Phone Corp. v. U.S.*, 238 F.2d 266 (D.C. Cir. 1956).

allowed private companies to use microwave technology to haul their own private long-distance communications. Until this time, not even private companies could operate their own networks; they had to rely on the AT&T network almost exclusively. In 1964, MCI applied to the FCC for authorization to construct and operate a microwave network to carry other private companies' long-distance traffic between two United States cities. MCI subsequently sought FCC authority to carry all types of long-distance traffic and to interconnect with AT&T's network using this network [1]. The FCC initially determined that these activities should not be authorized on the ground that it was not in the public interest. MCI subsequently appealed the decision to a federal court, which overturned the FCC's decision and ordered the FCC to issue to MCI an authorization to provide a full range of interstate services and interconnect with AT&T's public switched network. This was a major step in opening the door to telecommunications competition in the United States.

With some competition in the long-distance industry now authorized, AT&T began to engage in what the U.S. government would come to view as anticompetitive practices in order to protect its existing market share. In response, in 1974, the U.S. Department of Justice (DoJ) filed an antitrust suit against AT&T to halt its anticompetitive practices. In 1982, AT&T and the DoJ negotiated a consent decree to settle this antitrust suit. This decision became known as the Modified Final Judgment (MFJ) and resulted in AT&T divesting itself of its 22 local phone companies [2]. This left AT&T as (in terms of its service arm) primarily a domestic and long-distance service provider. The divested companies were specifically prohibited until the late 1990s (pursuant to the Telecommunications Act of 1996) to provide long-distance services, as well as from entering certain other lines of business.

Because of these developments, the U.S. long-distance market transformed from a market of fledging competitors overshadowed by a single dominant service provider to a market characterized by a declining market share for AT&T (but, as is often the case with monopolies that become competitive, with increasing revenues) and significant growth opportunities for competitors. As a result, today U.S. consumers (both business and residential) have a wide choice of long-distance service providers available to them at competitive prices. More recently, in the past few years, competition has begun to blossom in the U.S. local services market as well.

Shortly after the United States began to open its long-distance and other telecommunications service markets to competition, other countries also began to take steps in this direction. One of the first and most successful was the United Kingdom. However, unlike the United States, competition in the United Kingdom was not primarily an industry-led endeavor but a cornerstone of the then-Thatcher government's reform plan. In 1981, the United Kingdom, pursuant to the newly enacted British Telecommunications Act of 1981, separated the telecommunications functions of its monopoly service provider from its postal entity, creating British Telecommunications and an independent regulator [3]. The law also expressly authorized facilities-based and resale competition to British Telecommunications. Shortly thereafter, in response to the newly enacted law in 1982, Mercury was issued the first competitive facilities-based authorization in the United Kingdom. During this period, the United Kingdom began reducing its ownership in British Telecommunications, ultimately eliminating government ownership [4].

The subsidiary goals of the United Kingdom process were twofold: to encourage the telecommunications service sector and to reverse the decline of British technological leadership, primarily in the equipment market [5]. While the process did not fulfill the second goal, the first goal was fully met. Today, the United Kingdom has vibrant competition in the telecommunications service sector. This result has benefited British Telecommunications, which had to adapt to a competitive market in order to survive and is now more profitable than ever;[10] business consumers, who have been able to take advantage of low international and other service rates; and many of the new entrant service providers, which have been able to capture valuable market share.[11] In addition, residential consumers in general have benefited, because of the increase in build-out of the telecommunications infrastructure, generally lower rates, the increased choice in service providers, and the technological innovation that the new competition has propelled.

Other key country markets soon followed, mostly pursuing the United Kingdom's first goal—to encourage the telecommunications service sector—but often having subsidiary goals, such as improving the overall health of the economy by attracting new business and providing improved access to the telecommunications network. Most notably, the

10. By March 2001, British Telecommunications had completed $28,930.8 million in business.

11. For example, the United Kingdom has one of the most advanced mobile service markets in the world and was one of the first countries to auction spectrum for advanced mobile service operations.

European Union, capitalizing on the success of the U.K. market, began its own plans to allow competition in the entire European Union. The first major in-road was the 1987 Green Paper, which initiated a series of legislative actions in the European Union to open the telecommunications market and to harmonize the regulatory frameworks of EU member states [6].

Subsequently, in its 1988 Action Plan, the EU Commission set forth the EU's liberalization plans, including the opening of the telecommunications terminal market and unrestricted recognition of type approval for terminal equipment; the progressive opening of the telecommunications services market to competition; the separation of the regulatory and operational activities of the incumbent monopoly services providers; the establishment of conditions ensuring open access to network and services; and application of the EU's competition rules to the telecommunications sector [6]. The European Union implemented its action plan over the next decade through EU Commission directives and other EU rulings. Subsequently, almost all EU member states were required to bring their own telecommunications markets in line with the European Union requirements by 1998.

Other regions of the world also followed suit. Chile was one of the early market openers in South America. Specifically, in the early 1990s, in response to the changing political regime in Chile, the government began to permit competition in many of its key sectoral markets, including in its telecommunications market. The Chilean telecommunications sector was the first to be privatized in South America and is one of the most liberal telecommunications markets in South America. Specifically, the Chilean government finalized the privatization process in 1989 when it completed the sale of all state-owned telecommunications companies, including the two dominant service providers, Compañia de Teléfonos de Chile (CTC, domestic service) and Empresa Nacional de Telecomunicaciones (ENTEL, long-distance service). Today, all service providers in Chile are privately owned. The Chilean telecommunications sector is completely open to investment by local Chilean and foreign companies. An independent regulator, the Undersecretariat of Telecommunications (SUBTEL), is the primary authority for supervising and regulating all telecommunications services.[12]

12. SUBTEL is responsible for issuing all franchises, licenses, and concessions for services; establishing technical standards for equipment; supervising the country's operating companies; and developing national and international telecommunications policies. SUBTEL also has responsibility for the allocation of radio frequencies.

Further, today the effects of competition can also be seen in many global markets. For example, as of July 2000, more than 2,800 companies were authorized to build international telephone networks. In 1997, that number was less than 600. It has been estimated that in some markets, such as Germany and Hong Kong, new entrants were able to gain over a third of international minutes in just 1 year's time [7].

However, as discussed, competition is still only beginning in many developing countries—many of these markets are still either completely closed, such as Vietnam, Jamaica, and Iran, or allow only limited amounts of managed competition, such as Egypt or the United Arab Emirates. As discussed further below, because of the lack of competition in these markets, they are plagued with poor infrastructures, high rates for service, and generally limited access to the network in most rural and remote regions. Further, other sectoral industrial players are often hesitant to enter these markets, since they will not be able to have reliable communications. If businesses do enter these markets, they may seek to have their communications services provided by alternative, and sometimes illegal or "gray," methods. In the most closed of these markets, it can be estimated that it may take as long as a decade until the concept of competition is introduced in a meaningful manner.

2.2.2 Influences

In order to understand why competition is today the desired result of most governments in regulating their telecommunications markets, one must comprehend the influences that have helped shape this goal. There are numerous influences that have assisted in leading to this result, with the key ones including the needs of the business community, technological innovations, government needs, societal needs, and international pressure. Below is a brief discussion of how each of these influences have led governments to seek competition in their telecommunications market.

2.2.3 The needs of the business community

Perhaps one of the biggest influences in creating competition in the telecommunications market is the effort of the business community. There are two main constituent business groups active in telecommunications. First, the most prominent and outspoken group is the service providers, who wish to enter a new market or who are already operating in a market. The second, less vocal, equally important, group is the business consumers of telecommunications services.

Both constituent groups will often make the government well aware of their needs through advocacy either on a stand-alone basis or together in larger "industry groups." For example, many service providers and business customers participate in local chambers of commerce. Often, these constituencies, even if competitors, have similar goals and can effectively lobby a single position as a mass group, thus leveraging numbers to garner more political support and clout.

Service providers

Competitive service providers recognize that competition is good business. Specifically, these companies recognize that in order to succeed, a level playing field must exist for all providers. Such goals are pursued by start-ups, such as Global Crossing, as well as by former telecommunications monopolies, such as Deutsche Telekom. Accordingly, most efforts by these providers are aimed at the government adopting the basic regulatory formula. Across the board, competitive service providers recognize that often the incumbents become stronger companies by becoming more efficient, more innovative, quicker to market, and more responsive to their customers. These interests of service providers will be discussed in greater detail in subsequent chapters.

Business consumers

More and more companies are operating on a global basis, with physical locations in many countries and their customer reach extending to almost any place on earth. As companies continue to expand their reach, reliable telecommunications services are often an important key to these companies' success. Accordingly, companies often will consider telecommunications services, in terms of access, price, and reliability, in determining where to locate their international offices and which markets to engage in e-commerce. Countries may lose out on business opportunities if their telecommunications infrastructure and market availability are not consistent with the high-end standards that these companies are seeking.

In addition, business customers are seeking high-quality, high-bandwidth services around the world. No longer are businesses willing to have segmented networks. On the contrary, businesses expect to have the ability to have access to a global network with the same level and quality of service whether they are operating in Calcutta, New York, or Seoul. To this end, business customers often demand end-to-end management of global, voice, data, and Internet services with a single point of contact through a global service provider.

Accordingly, to capture these goals, business consumers often will seek to locate their business operations in countries where there is a competitive telecommunications market. This enables them to obtain the following goals: the largest choice of service providers; lower prices; an improved infrastructure; and access to reliable, innovative, and efficient telecommunications services.

2.2.4 Technological innovations

Another key influence on the creation of a competitive telecommunications market is emerging technology. New technologies and applications are shrinking the globe and eliminating borders. These technologies, such as the Internet and advanced mobile telephony applications, allow consumers greater access to communications and often require greater bandwidth than traditional wireline telephony.

As more Internet use and applications are developed, the demand for additional bandwidth will continue to place tremendous pressure on existing networks that are already overutilized and undersized. For example, in 2000, as some Internet service providers began to upgrade their international backbone connections from 155 Mbps to 2.5 Gbps, there was a tripling or even quadrupling of bandwidth on many international routes [8]. Similarly, the increased capacity put a strain on existing domestic infrastructures.

Another huge impact on traffic flows is the increase in mobile use.[13] Traffic flows will continue to increase exponentially as third-generation mobile telephony technology is deployed.[14] Once again, this increased use puts tremendous pressure on existing telecommunications networks.

Accordingly, existing telecommunications infrastructure is often insufficient to handle these new demands on capacity. Governments must actively seek new ways to meet their growing infrastructure requirements or face missing out on the ability to capture the benefits of the information age for their citizens and as a revenue provider. Competition is often

13. For example, in 1999, approximately 11.5% of international calls were placed from mobile telephones [8].

14. Third-generation mobile service is the next generation of advanced wireless telecommunications systems. These systems will offer higher data speeds and increased capacity to support fixed and mobile service on wireless devices, including small pocket terminals, handheld telephones, and laptop computers.

the only way to ensure that there are sufficient resources available in a marketplace to meet these demands.

2.2.5 Government needs

Many governments have also recently recognized that they cannot operate in a vacuum; they must open their markets to international competition in order to be successful both in terms of bringing in new or retaining current business, as well as in order to provide needed telecommunications services to their citizens.

Despite these needs, many governments are also often faced with similarly strong political pressure to retain the status quo. This may be caused by many factors, including national pride stemming from the national service provider, many monopoly service providers are large employers, monopoly providers are often a cash cow for the government, and the government is hesitant to allow foreign investors into its market for the provision of what is seen as an essential service. Furthermore, many of the world's duopoly markets have a socialist political philosophy, at least in part. This may make implementing the components of competition, such as privatization, politically difficult and at odds with the basic tenets of the country's political structure. Accordingly, governments often have to engage in a balancing act in the timing and process for opening their telecommunications markets to competition in order to have the political support that is necessary for a successful market opening.[15]

2.2.6 Societal needs

Consumers are often underrepresented in the formation of telecommunications regulation. Although many governing laws for telecommunications regulatory schemes state that a goal of the law is to serve the public interest, it is often an unclear goal in terms of achievement, in large part because of the monies required to meet this goal. Accordingly, governments often have to grapple with how to best achieve this goal.

In general, consumers have several major concerns, which include access to services, rate levels, the ability to receive service, and the quality of services. Many citizens do not even have access to make a basic phone call, much less have the ability to access new technologies. Furthermore, often the cost of telecommunications services is prohibitive to a large portion of a country's citizenry. Of equal importance is that telecommunica-

---

15. This balancing process must also take into account the specific form of legal regime.

tions services are often unreliable. For example, in areas of Eastern Europe, such as Romania, portions of the existing infrastructure of incumbent service providers are so old and decrepit that in severe rainstorms outages occur for days on end. While governments are often aware of these situations, traditionally they have lacked the political power from consumers that is needed to react; this situation may be a lower political priority than reacting, for example, to the more vocal military or public health requirements. This prioritization is beginning to change as more and more countries are beginning to adopt the American form of citizen advocacy and as populations have become more sophisticated about communications choices.

In the United States, consumers are generally represented through the lobbying efforts of public interest groups, such as the Consumer Federation of America. As competition enters more and more markets, similar consumer advocate groups are also beginning to blossom around the world. For example, in Brussels, the seat of the EU, there are often visits to lobby the EU on behalf of consumer groups from both member states and nonmember states. Another important voice consumers have is through the voting booth. In the 1990s and 2000, quite a few candidates for high offices around the world have included telecommunications reform as an important plank of their campaigns. For example, in Romania, an issue in the 2000 presidential campaign included the prices for telecommunications services, which had risen dramatically since the privatization of the incumbent service provider, Romtelecomm.

2.2.7 International pressure

A very important influence on the creation of competitive markets is international pressure. International pressure can take varying forms—from that imposed by formal international agreements and treaties to pressure exerted by international organizations and informal pressure by other governments.

Perhaps the largest most recent influence has been the adoption of the World Trade Organization's Agreement on Basic Telecommunications (WTO Agreement), which became effective on February 5, 1998 [9]. As discussed, the WTO Agreement sets the broad framework for the creation of a competitive market in telecommunications services through a trade in service treaty. The WTO Agreement currently includes commitments from over 72 member states, covering a majority of the global telecommunications service market. These commitments cover market access and

TABLE 2.1 Overview of WTO Offers

REGION	COMMENT
Europe	Generally significant
Latin America	Somewhat positive
Asia & Africa	Varies widely
Non-WTO	Key markets include Russia, Vietnam, and Saudi Arabia

national treatment, foreign investment, and dispute settlement. Many member states also signed on the Reference Paper, which provides for key regulatory principles to which the member states will abide. In addition, the WTO Agreement contains a formal dispute settlement mechanism.

Although the WTO Agreement created a very good framework for market opening, it is far from perfect. There are many market realities associated with the WTO Agreement. First, as detailed in Table 2.1, although many countries committed to grant market access, the dates and terms of these commitments vary. Many other countries, although members of the WTO, have made no commitments. These countries, however, are able to reap the benefits of the other commitments without committing to anything more than the basic WTO principle of national treatment. Further, many key markets, including India and Saudi Arabia, are not members of the WTO. Accordingly, the battle for market access and the creation of a competitive telecommunications market is still on a country-by-country basis, although the WTO Agreement does provide market openers important leverage in countries where there are meaningful commitments.

Another major international pressure point was the European Union's decision to open its telecommunications market to competition in 1998. Specifically, as discussed earlier, the EU mandated that most of its member states must implement a competitive telecommunications regime by 1998. Once the European Union began its liberalization process in earnest, other Western European countries began to follow suit and, recently, many Eastern European countries that are seeking to gain entry into the European Union as member states have begun to implement EU directives. In many respects, the movement by the EU laid the groundwork for the successful conclusion of the WTO Agreement.

Other important groups include the International Telecommunication Union's Telecommunications Sector and the Organization for Economic and Cooperative Development, as well as regional groups such as the Inter-American Telecommunications Commission and Asia Pacific Telecommunications group. In these forums, policy discussions are regularly held on competition and how best to liberalize, privatize, and regulate telecommunications markets. Originally, many of these organizations were not very sophisticated in regard to these issues, primarily because government-owned monopolies controlled the course of the meetings. As more and more governments open their markets to competition, and as the private sector players increase their active participation in these organizations, these antiquated viewpoints will no longer control the course of the policy discussions. Over the next several years, one can expect greater activities in this area that directly impact the formulation of telecommunications regulatory regimes.

2.2.8 Subsidiary goals

The primary reason that governments create a competitive telecommunications market is to obtain their subsidiary goals. Subsidiary goals commonly fall into two classes: direct and indirect goals. Direct goals are those where there is a linear relation to the creation of competition. Such goals can include achieving lower prices for consumers, creating greater consumer choice in terms of service providers and services available, the introduction of innovative technologies, and increasing the flexibility of service offerings.

These direct goals are almost universally recognized when a market becomes competitive. For example, reduced rates for services have been widespread. In Western Europe alone, the rates for international calls have dropped up to 90% in as little as 2 years [7].

Indirect goals, on the other hand, are those goals that do not directly flow from the creation of a competitive telecommunications market. These goals can include attracting new business or attempting to retain old business in nontelecommunications fields, such as banking, through the creation of a modern and reliable telecommunications infrastructure, providing increased services to the public in terms of health or education through the introduction of tele-medicine and tele-education, and encouraging the development of the high-technology sector.

As discussed below, subsidiary goals are often the driving force behind liberalization and privatization processes. In fact, many modern telecommunications laws specifically provide that the goals of telecommunications regulation are exactly these subsidiary goals. For example,

Venezuela's recently enacted telecommunications law provides for the following [10]:

1. To defend the interests of users, assuring their right to access to the telecommunications services in appropriate quality conditions and to protect, upon rendering such services, the effectiveness of the constitutional rights, especially those regarding respect for honor, privacy, secrecy in communications, and protection to the youth and the children. To that end, obligations may be imposed on the operators of the services in order to assure such rights.
2. To promote and help the exercise of the people's right to establish means of community sound broadcasting and open television system of public service in order for the exercise of a free and plural communication.
3. To procure competition conditions among the service operators.
4. To promote the development and use of new services, networks, and technologies, when available, and access to them in equal conditions of persons and further the geographic integration of space and the economic and social cohesion.
5. To further the efficient integration of telecommunications services.
6. To promote the research, development, and transfer of telecommunications technology, as well as training and employment in the area.
7. To render possible the efficient, effective, and undisturbed use of the limited telecommunications resources, such as the numeration and the radio spectrum, as well as the appropriate protection of the latter.
8. To incorporate and assure compliance with the obligations of universal service, quality, and standard minimum coverage goals and the obligations related to security and defense regarding telecommunications.
9. To favor the harmonious development of the telecommunications systems in the geographical space according to the law.

10. To favor development of the regional integration mechanisms to which the Republic is a party and promote the participation of the country in international telecommunications agencies.
11. To promote national and international investment for the modernization and development of the telecommunications sector.

Therefore, the goals of Venezuela in creating a competitive telecommunications market expressly include the creation of competition in the telecommunications market, as well as creating new services, increasing the level of technology, ensuring universal service, and other similar goals.

2.3 Privatization

Often the first step toward the creation of a competitive telecommunications market is the privatization of the incumbent government-owned and -operated monopoly service provider. However, this is often a very difficult first step for governments to take. For many governments the monopoly service provider is a cash cow and a bloated bureaucracy. In addition, it provides the government with direct control over the telecommunications network. Accordingly, it is difficult for governments from both a political and economic viewpoint to surrender this important source of revenue and jobs. However, governments that either delay the privatization process or do not proceed with it because of concerns over these issues are shortsighted and will likely be negatively impacted in future development in the telecommunications sector, as well as in other industries. This approach neglects to take into account that an influx of capital and expertise can be brought in through the effective privatization of the incumbent provider. As discussed below, this can lead to an improved telecommunications infrastructure and the introduction of innovative services to consumers.

Further, there are many other benefits that can be captured through the privatization process for governments, including, but not limited to, the introduction of lower rates through improved efficiency in operations and better customer service. This also includes laying the basis for a more competitive telecommunications market to be developed through the additions of liberalization and effective regulation.

One of the most financially attractive draws of the privatization process is the direct economic contributions that may flow into the national treasury by investors. This money may be allocated by the government to assist in the development of other sectors or to achieve other government goals, such as improving medical care or building highways.

Another financial benefit is the attraction of foreign capital to improve the existing telecommunications infrastructure. The government may also be able to obtain significant commitments from the investor in terms of build-out of infrastructure or the provision of new services. Such services would often not be provided if the incumbent service provider did not receive an infusion of capital. Further, through privatization the incumbent provider may be able to obtain expertise in the technical, marketing, and other areas in which the incumbent provider is lacking.

In addition, if the ultimate goal of the government in its telecommunications market is the creation of a competitive market, even in a limited form, it is imperative that the government remove itself from ownership and operation of the incumbent service provider. If the government retains an ownership interest, except perhaps in the most limited form,[16] it will have the incentive (or there will be the appearance of the ability) to discriminate in favor of the incumbent service provider in the market. Accordingly, privatization is an essential element to the creation of a competitive market and to ensure that the additional components of the basic regulatory equation, liberalization and deregulation, are also met.

2.3.1 Different forms of privatization

As discussed in more detail later, the form of privatization of the incumbent telecommunications service providers that a country takes may be different in each market. Specifically, the major forms include public stock offerings, direct investment in the incumbent service provider, and the formation of a joint venture.

The most common form of privatization of the incumbent telecommunications provider is the government sale of a strategic and controlling share of the incumbent monopoly service provider. This process is generally done through an auction process. This auction process will usually take one of two forms. First, it may be a straight auction process, where the contestants must meet minimum qualifications to participate, and, once they meet those, the auction will be determined based on the monetary level of the bid. A second type of auction is one that is not based

16. For example, some governments retain a small golden share in the incumbent service provider.

solely on money but also takes into account commitments that the investor is willing to make if it wins the auction. These commitments can include such things as investment in infrastructure, build-out, and the like. Another possible process is a beauty contest. In a beauty contest, each applicant will submit information on its qualifications as well as what commitments it is willing to make if it is the chosen investor.

Both the beauty contest and the combined auction/beauty contest approaches are fairly subjective and may lead to allegations of bias in the process, depending upon the result. Many governments use this rationale for pursuing a straightforward financial auction. These governments are then able to recognize the larger monetary rewards that generally accompany this approach, even though there is a cost in terms of achieving some of their subsidiary goals.

By using any of these processes, there may be limits imposed on the strategic investor. One of the most common forms of limits is that on foreign investment. Another limit is often the amount of ownership that the investor may hold in the service provider as a whole. Often, the government will only privatize a portion of the service provider and will hold the remaining share, which will later be released in the stock market or otherwise sold off.

Limited ownership rarely provides investors with the same incentives to invest resources. Further, this form of privatization may not always permit the investor to obtain management control. Without management control, there may be difficulties in achieving many of the benefits that typically accompany privatizations, such as innovative management, the introduction of new services, and increased investment, among other things.

2.3.2 The status of the privatization process

Most of the major markets around the globe have privatized their incumbent service providers. For example, all of Western Europe, almost all of the Americas, and significant markets of Eastern Europe and Asia have gone through the privatization process. Africa, however, still needs to make great strides in this area.[17] There are also other key markets where privatization has yet to occur. These tend to include countries that still have fairly closed economies, such as Thailand, Saudi Arabia, and China,

17. Only a handful of African nations have even partially privatized their incumbent service providers. These include Ghana and South Africa.

or countries where many of the inhabitants are employed by the telecommunications sector, such as Costa Rica.[18] In addition, since the telecommunications market became more glutted with opportunities, some countries, such as Ecuador, have held off on moving forward with the privatization process until they feel that they can recognize an appropriate value for their service provider. It can be anticipated that in the next decade all but a few stragglers will have completed the full or partial privatization of their government-owned and government-operated incumbent service providers.

2.4 Liberalization

Liberalization in its most basic form is the creation of actual competition in a specific sector of the telecommunications market, such that new entrants are allowed unfettered entry. The most dramatic effects of liberalization began to occur in 1998, as discussed below, the year the EU liberalization effort was implemented fully and the year the WTO Agreement on Basic Telecommunications Services came into effect. However, certain markets still refuse to surrender to the growing trend of liberalization. A market characterized by a single company operating as a monopoly is often seen as easier to "regulate" by the government than where there are numerous companies operating in multiple market segments characterized by direct competition. Despite these holdouts, many experts believe that most markets will succumb to the pressure to liberalize their telecommunications market in the next few years.

2.4.1 The components of a successful liberalization process

There are at least four basic components that are key to the success of the liberalization process. These are as follows:

1. The government must instill confidence in the potential service providers so that they will want to enter the market. In order to achieve this, a clear set of licensing rules and procedures must be issued. This will allow entities wishing to obtain authorizations to understand how the process works and with what requirements they are required to comply.

18. Since Costa Rica does not have a military, many of its citizens find employment with the incumbent monopoly telecommunications service providers.

2. The process that is established must be fair and transparent. Accordingly, applicants must understand the process for going forward, whether it be an auction, beauty contest, straight application, or a combination, and a reasoned decision, based on existing law and regulation, must be provided for any actions.
3. Licensing rules must be clear. Applicants must be aware of what is required of them to obtain and retain a regulatory authorization.
4. An appeals process that ultimately allows access to the judiciary should be established so that an applicant that is rejected or receives an authorization with conditions that it believes are unwarranted may seek redress from a neutral third party. Further, this appeals process must allow access to aggrieved parties from subsequent regulatory action or inaction.

2.4.2 Liberalization may be phased in

Telecommunications liberalization is a new phenomenon in many countries. Accordingly, in many instances, the liberalization process often takes place in phases. Governments may structure their markets to open only limited services, such as Internet or e-mail to competition, or introduce new services, such as mobile services, on a competitive basis, while retaining the existing monopoly of limited amount of competition for more traditional services.[19]

As discussed, often governments will utilize a limited form of market access as a test to see how competition works or does not work in their telecommunications market.[20] For example, many governments only initially allow competition in the value-added telecommunications sector for services such as Internet, high-speed data, and private voice networks. This approach serves several interests. First, it provides businesses with access to new service providers for many of the applications about which they are

19. Argentina is a case in point. Until November 2000, it only allowed competition in its domestic services market.
20. In the United States, for example, the FCC, as early as the 1970s, began fostering a competitive market for enhanced or value-added services. See *Regulations and Policy Problems Presented by the Interdependence of Computer and Communications Services and Facilities.* 28 F.C.C. 2d 291 (1970) (tentative decision); 28 F.C.C. 2d 267 (1971) (final decision), *aff'd in part sub nom; GTE Service Corp. v FCC*, 474 F.2d 724 (2d Cir 1972), *aff'd by order* 40 FCC 2d 293 (1973).

most concerned. This may allow the government to retain existing business or attract new businesses to the market. Second, introducing only limited competition allows the government to continue to benefit from either its incumbent government-owned basic service provider or leverage off of the privatization process by providing an exclusive period of time for the provision of basic services. In the case of a government-owned monopoly, the government will be able to continue to recognize the revenue associated with that operation. In terms of the exclusivity for a privatized basic service provider, the government will either be able to meet its commitment to the investors or possibly even obtain new commitments by keeping new competition at bay for a longer period of time. While this approach may bring certain immediate benefits, over the long term, as markets such as Singapore have recognized,[21] limited forms of competition often stifle the realization of longer-term goals, such as attracting new businesses, achieving lower rates, and increasing service penetration levels.

Another manner in which competition may be limited is through the imposition of strict foreign ownership requirements. Many countries, at least initially, only allow a limited form of foreign ownership in new entrants. Accordingly, foreign investors must team with local service providers and other investors in order to obtain a regulatory authorization. Often, this will result in less new entrants than in a market where 100% foreign ownership is allowed, since limited investment may mean limited control of the operating entity (this, as discussed subsequently, will depend in part on the structure of the entity). Limited control will often dissuade investment. However, many governments, as discussed earlier, are hesitant to give control of what is still seen as a strategically important resource to foreign entities. Often, this hesitancy is politically motivated. Nonetheless, it is likely that as the world becomes increasingly global with the need for increased capital in order to meet infrastructure requirements, we will see foreign investment limits on telecommunications licensees further diminish.

21. Singapore originally was planning to have a duolopoly market for basic telecommunications services for a set period of time. However, at the time that the duopoly was to begin, the Singapore government recognized that failure to initiate full competition would lead to the loss of a tremendous opportunity to attract and retain businesses in other sectors. Accordingly, the Singapore government, much to the surprise of global service providers, fully opened its doors to competition instead of limited competition.

2.4.3 The effectiveness of the liberalization process

The effectiveness of the liberalization process can be impacted by many factors, including timing, the shape of the economy, the acceptance of competition by the citizenry, and the government's commitment to liberalization. In most countries, however, it is the government's commitment to liberalization that has the most direct impact on the success of the process.

A brief overview of the world demonstrates that many countries have successfully liberalized their telecommunications markets. For example, the United States, most of the EU member states, and Australia have successfully fully liberalized their markets. Many other countries, such as the larger economies of South America (including Argentina and Venezuela) and some key Asian economies (such as Japan and Hong Kong) are well on their way. Most of these countries have been able to leverage the political will necessary to liberalize their telecommunications markets.

Despite these successes, there are many other markets, such as South Africa and South Korea (where the liberalization process is underway), where a successful liberalization process is not assured. Often, this is because of an unclear process and the failure of the government to institute an effective regulatory regime. This latter point will be discussed below.

Many other markets have not begun the liberalization process or have only made tiny inroads. These markets include the countries of sub-Saharan Africa, Saudi Arabia, the United Arab Emirates, Egypt, Thailand, and Vietnam, among many others. In these cases, as we will explore throughout this book, the government has not had the political will to move forward fully or at all with the liberalization effort.

2.4.4 The result

Institution of a successful liberalization process has led to three major results: the emergence of global service providers, dramatic price cuts, and new service offerings and improved infrastructure. These results will be discussed below.

The emergence of global service providers

A key result of the liberalization process has been the emergence of global service providers. Initially, service providers were constrained by regulation to operate solely or primarily in their home markets. However, as new opportunities opened up through the liberalization and privatization processes, the larger, well-financed service providers, such as AT&T, British Telecommunications, and Telefonica, were able to move into new markets outside of their home markets as service providers. In this capacity, these providers are able to capture synergies of their international operations, often choosing to initiate operations in markets where their largest customers are or where the most international traffic from their originating country terminates.

An interesting note is that at least in terms of the international long-distance market, it appears that markets that liberalized more recently, such as Germany, are seeing more dramatic increases in competition than in markets, such as the United States or the United Kingdom, that liberalized that market substantially earlier. For example, as discussed earlier, the United States introduced competition in the international long-distance market in the early 1980s. As of 1995, AT&T still held a substantial percentage of market share. Germany, on the other hand, introduced competition in 1998 and as early as 1 year later, its incumbent's market share had fallen to almost half and is continuing to decline. This dramatic difference may be traced to the emergence of global service providers in the mid to late 1990s that operate on a multinational or global basis and already have the experience to be successful in newly competitive markets. These service providers include WorldCom, Inc., Telefonica, Teleglobe, World Access, and Global One. This is in contrast to the markets that were among the first to open to competition, where the new entrants were primarily start-ups with little or no experience in telecommunications or in entering new markets.

Global service providers hold a number of advantages over other new entrants in establishing themselves in new telecommunications markets. First, they tend to have access to significant amounts of cash. Second, they are able to leverage their existing skills, technology, networks, resources (such as marketing) in all the countries in which they operate. Third, many global service providers provide services to global customers. Accordingly, they are able to leverage these customers' global presence as they enter new markets. Another significant advantage these companies have is that they have proven themselves as successful service providers. Therefore, governments are likely to put stock in their track records, and new customers are less likely to question their ability to provide the services they promise in the new market. As more and more markets liberalize, we can expect to see an increase in the number and the impact of the global service providers.

Dramatic price cuts

Another direct result of liberalization has been dramatic price cuts for consumers. In some markets, the direct result of liberalization of the market has been the entry by service providers that are willing to operate at a loss or near thereto in order to gain market share. In this regard, it is important to note that there may be no way to make up the loss of revenue, because demand has not necessarily kept pace with price reductions. For example, in Germany, prices in some routes have dropped upwards of 90%, but call volumes only grew by 45% [11].

An important issue that this raises is how the incumbent service provider responds. If the incumbent reduces its prices, its business case protections may fall off and, hence, its profitability. However, if it fails to react, losing customers may similarly disadvantage it. Regulation may assist in curbing the behavior of the incumbent in these situations, so that competitors are not harmed in their efforts to enter the market and consumers benefit in the long run.

Additionally, price cuts may begin because of increased efficiencies in operations. For example, the introduction of competition may mean that global service providers are able to reduce costs through the introduction of regional networks (such as in Eastern Europe) or countrywide networks. Further, new entrants may be able to use newer, cheaper technology than what previously existed in the market. This may allow the competitive services to be priced lower than previously existed in the market. Another cost savings may come from the ability to build new infrastructure, which was previously the domain of the incumbent service provider. Global Crossing has been extremely successful in this arena, by building modern and reliable transoceanic cables.

New service offerings and improved infrastructure

Another area that benefits from market liberalization is the introduction of new service offerings and improved infrastructure. It is important to remember that in a regime with only one service provider, it is often difficult for that provider to be able to offer all types of services. Accordingly, the incumbent service provider, during its phase as a monopoly or near-monopoly, will concentrate on basic telecommunications services, such as local and long-distance services. Such a focus will often lead them to fail to adequately provide new services, such as Frame Relay or Asynchronous Transfer Mode, or build out their infrastructure adequately to meet increased demand for service.

The introduction of a liberalized market will often stimulate the introduction of novel services. Many new entrants will attempt to focus on sectors of the telecommunications market that were previously not provided, either adequately or at all. Furthermore, many of these new services will also require an improved infrastructure. New service customers, or the business customers of the global service providers, will often find the existing reliability and capacity of the incumbent's network unsatisfactory. Accordingly, a focus of many new entrants will include improving the existing infrastructure of the liberalizing telecommunications market.

2.5 An effective regulatory regime

In order to ensure that competition is able to flourish in the telecommunications marketplace, it is imperative that a viable regulatory framework be established. At a minimum, this framework must include:

- A fair and objective licensing process;
- Access to scarce resources, such as spectrum and rights of way;
- Effective regulations (requiring an independent regulator willing to enforce its regulations in a meaningful manner);[22]
- Transparency in processes;
- Competitive (including dominant service provider) safeguards, including nondiscriminatory, cost-based interconnection;
- Unbundled network elements and access to rights of way.

These factors will be discussed in further detail throughout this book. Below we focus on the political realities of establishing this type of regime.

It is quite often politically difficult for an effective regulatory regime to be established. First, it takes political will to isolate the regulatory structure from political considerations. Further, the incumbent service provider, which still may retain strong ties to the government, may wo against the creation of such a structure, since it is likely to be the targ

22. An independent regulator makes decisions based on expert judgment wit influence from the executive or legislative branches.

the appropriate government authorities. This is different from regimes, such as the United States, where the regulator, the FCC, obtains its primary enforcement authority from the Communications Act of 1934 and its regulations, not from its licensing authority. Accordingly, oversight by the FCC occurs primarily through day-to-day regulation.

However, while there may not be a single successful form for creating an effective regulatory regime, competition in a marketplace will not exist in a meaningful manner without truly effective regulations. Many examples of this abound. For instance, in Mexico, failure of the Mexican government to constrain the anticompetitive behavior of Telmex, the incumbent service provider, has seriously impeded the ability of new competitors to capture market share in the Mexican telecommunications market. Further, many resources that could be invested in building out improved infrastructure and introducing new services instead have been utilized in court cases and lobbying in an effort to have effective regulation of Telmex.

2.6 Deregulation: a hidden component

One of the most interesting dynamics as competition comes into the marketplace is the need for more, not less, regulation. Hence, it is a misnomer when liberalization and deregulation are used interchangeably. As Figure 2.1 shows, deregulation generally only occurs once a competitive market has been firmly established, and the opportunity for anticompetitive abuses has been limited.[23] This is a rare instance and currently tends to happen only in limited market sectors.

An important concept to understand when speaking of deregulation is that once deregulation occurs, there is no effective method in which to curtail anticompetitive abuses or ensure that a level playing field for service providers exists. In addition, the ability to protect consumers from abuses in the market is often limited. Accordingly, many governments are

23. An interesting exception to this general rule was the case of New Zealand. Initially, New Zealand fully liberalized its market and even took the bold step of not establishing a regulator. The government instead left the antitrust commission with the ability to police potential anticompetitive abuses. However, since Telecom New Zealand remained dominant, without any real regulatory constraints, it was able to dominate the marketplace, hence hindering the development of a truly competitive market.

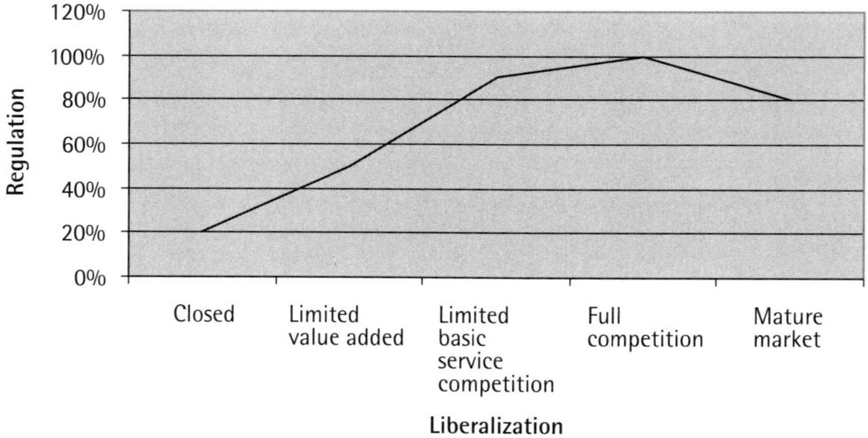

Figure 2.1 Deregulation chart.

hesitant to allow deregulation at a point that is too early in the liberalization process.

2.7 The result: recent case studies

2.7.1 Switzerland: a non-EU country that has adopted EU-like telecommunications regulations

Switzerland has traditionally been a country that has resisted entry into the European Union in an effort to retain its autonomy (although Switzerland is an EFTA member). However, in the 1990s it suffered economic stagnation and began to look for ways to improve the economy. One key method for achieving this was the passage of its 1998 Telecommunications Law, which supplemented the establishment in 1992 of an independent regulator for telecommunications. Specifically, the 1998 law resulted in the following key events:

- Splitting the telecommunications from the postal function of the Swiss postal and telecommunications entity;
- Setting the groundwork for the privatization of Swisscom (the former monopoly telephone company);
- Expanding the role of the regulator;

- Introducing competition into the Swiss telecommunications market [12].

The 1998 law is very similar to the laws passed in the EU. Further, subsequent actions of the Swiss legislature and the regulator have resulted in a regulatory regime, which, for the most part, mirrors the implementation of liberalization efforts in the EU. In this manner, Switzerland has been able to become a key market for telecommunications. Following Switzerland's lead, other EFTA nations have also moved forward with EU-type liberalization efforts.

2.7.2 Brazil: a market on the move

Brazil is the fifth largest country in the world and has the fifth largest telecommunications industry. For years, however, the constitutional, legal, and policy limits prohibited most private sector participation in the telecommunications market. However, recent reforms of the 1990s in the telecommunications sector, including the amendment of the constitution, the passage of new laws, the establishment of an independent regulatory authority, and the privatization of the incumbent government-controlled monopoly telephony companies, have attracted private investment, from both Brazilian and foreign enterprises, and are transforming the Brazilian telecommunications market into one truly based on competition.

The modern history of the Brazilian telecommunications sector started in 1972, when the government created, through the consolidation of local telephone operations, a virtual monopoly service provider through the creation of Telecomunicacoes Brasileiras S.A. (Telebras), which was controlled by the government. A subsidiary of Telebras was also formed, Epresa Brasileira de Telecomunicacoes S.A. (Embratel), to provide domestic and international long-distance services, among other services.

This model existed until the late 1980s. The greatest strides in the reform of the Brazilian telecommunications sector began in 1988, when certain value-added services, such as paging and trunking, were liberalized. However, simultaneously, the Brazilian law was amended to formalize the Telebras/Embratel monopolies. In this manner, competition was hindered, and global service providers were only able to enter the market in the most limited way.

In 1995, in part because of increasing pressure internationally and from customers and service providers, the Brazilian government took a large step toward reforming its telecommunications market through the passage of a constitutional amendment [13]. This change expressly authorized the government to grant concessions, permits, and authorizations for "public" and "other" telecommunications services to both public and private companies. In response to this amendment, the government first passed in 1996 the Minimal Law [14], which allowed the government to issue regulations opening the value-added services, limited services, and cellular and satellite service markets to competition. Numerous authorizations were awarded pursuant to the Minimal Law. This included licensing of mobile service providers for which the government earned large sums of money,[24] as well as numerous business service providers, such as MCI Communications Corp. and Global One.

Following on the heels of the Minimal Law, the Brazilian legislature passed the General Telecommunications Law in 1997 [15]. This law was the first clear step to liberalizing and privatizing the Brazilian telecommunications market, as well as establishing an effective regulatory regime (which included the requirement of establishing an independent regulator). The stated objective of the law is to guarantee all Brazilians access to telecommunications services under reasonable prices, terms, and conditions.

In response to the General Telecommunications Law, the initial step the government took was to privatize the Telebras system. At the time of the privatization of the Telebras system, the government owned approximately 21% of the total equity of Telebras and controlled the voting shares, while the rest of the company was held through publicly traded stock.

In order to best sell the companies, the Brazilian government restructured the Telebras system into three regional wireline companies (for domestic service), eight regional cellular phone companies operating in Band A, and Embratel [16]. In a competitive bid, the government sold all the companies to foreign investors on July 29, 1998, for approximately $19 billion.

Almost simultaneously with the bidding, the government also began to open the public telephony market to new entrants. The government of

24. For example, the winner of the Band B concession for the São Paulo metropolitan area was awarded to a consortium led by Bell South. This consortium paid nearly $1 billion more than the next highest bidder. Bidders highly valued these concessions, because of the pent-up demand they believed existed because of the historically poor telecommunications services provided by the Telebras system.

Brazil achieved this by issuing one mirror concession of each regional Telebras company and Embratel. This duopoly was scheduled to last 2 or 3 years, depending upon certain conditions. Full competition was scheduled to occur on January 1, 2002.

Today, the Brazilian telecommunications market is operating in a much-improved manner. While basic services are still provided on a duopoly basis, these services have improved dramatically, with the introduction of new services, extended reach, and improved infrastructure accompanying them. This has even included increased services in the remote areas of Brazil, such as the Amazon region. Further, the introduction of competition in the mobile telephony market has led to a dramatic increase in these services. Today, a large percentage of Brazilians are dependent on the use of their mobile telephones. Further, the influx of limited service providers has skyrocketed. Services, such as FRS and Internet, which were almost unheard of just a few years ago, are available at reasonable rates throughout most of the country and are increasingly in demand.

2.7.3 Hong Kong: making progress

Another interesting case study is Hong Kong. In 1992, Hong Kong began a comprehensive review of its telecommunications market, resulting in a policy paper on the telecommunications sector being issued in 1994. As a result of this policy review, the government decided to set forth the following plan:

1. Establish OFTA as the independent regulatory body;[25]
2. Authorize competition to Hong Kong Telephone Company (HKTC) (the incumbent monopoly service provider) for domestic services upon the expiration of its license in June 1995.

Accordingly, in July 1995, OFTA authorized four companies to provide fixed telecommunications network services, with a scheduled exclusivity for these companies until 1998 in order to give them a head start on

25. Information regarding Hong Kong was obtained from the Office of the Telecommunications Authority's (OFTA) official home page at http://www.ofta.gov.hk/t-fact.html. The Office of the Telecommunications Authority (OFTA) was established on July 1, 1993, as a new government department. OFTA's primary responsibility is to regulate Hong Kong's increasingly competitive telecommunications industry.

the market.[26] The government also opened other markets to competition. This included value-added services, such as high-speed data, fax, and private network services.[27]

In 1997, the Hong Kong government initiated another review of the telecommunications market. This effort led to the early end of the Hong Kong Telecommunications International (HKTI) monopoly in international services; the establishment of the Information Technology and Broadcasting Bureau to coordinate development of telecommunications, broadcasting, and information technology policies; and further market reviews that ultimately resulted in the revision of Hong Kong's telecommunications laws. Accordingly, in 1998, the government negotiated an early end to HKTI's exclusivity in international telecommunications services (which was scheduled to last until 2006).[28] In exchange, HKTI had its license modified to include additional services. In January 2000, the Hong Kong market opened for facilities-based international services.

Through these processes, Hong Kong has been able to improve its position as an Asian telecommunications gateway in terms of achieving lower costs, greater service choices, and improved access to services.[29] However, as Japan and Singapore improve their regulatory efforts, Hong Kong will have to further refine its regulatory framework as well. Specifically, it will need to ensure that its regulatory regime is effective in curbing

---

26. The four FTNS licensees were HKTC, Hutchinson Cinnybucatuibs Kunutedm Bew, T&T Hong Kong Limited, and New World Telephone Limited. The licensing of these four new companies followed the termination of Hong Kong Telephone Company Limited's (HKTC) monopoly for the provision of local telephone services. Each of the four companies has been issued licenses to provide local fixed telecommunications services in a competitive marketplace.
27. Such services also include local and international telephone, fax, data, ISDN, and Centrex.
28. Under this exclusive license, HKTI is solely responsible for providing certain external circuits and services. Such services include external public telephone, fax, data, and television connections. Additionally, HKTI provides telegram service on a domestic and international level. HKTI also provides international, private, and leased circuits, as well as shore-to-ship and ground-to-air communications from Hong Kong.
29. A wide range of facsimile and data communications services are available in Hong Kong. By March 1997, there were over 314,000 facsimile lines. The growth rate continues at about 2,000 lines per month. Public switched data network services, Integrated Services Digital Network (ISDN) services, and Asynchronous Transfer Mode (ATM) services are also available in Hong Kong.

the monopolistic tendencies of HKTI. In this regard, the government of Hong Kong will have to ensure that it has effective regulations if it wishes to create a truly competitive telecommunications market.

2.8 Conclusion

As we have seen, the aim to achieve a competitive telecommunications market is an important government goal for many reasons. However, in order to ensure a competitive telecommunications market, the governments must be committed to fully implement the three major components of the basic regulatory formula for the creation of competition:

Privatization + Liberalization + Effective Regulation

We have briefly delved into each of these components in this chapter. Throughout this book, these components will recur not only as individual concepts but as motivating forces for other actions.

References

[1] *Allocation of Frequencies above 890 MHz*, 27 FCC 359 (1959).

[2] *U.S. v. AT&T (The Modified Final Judgment)*, 552 F. Supp. 131 (D.C. Cir. 1982).

[3] United Kingdom Statute 1981, Chapter 38 Pt I s 1: British Telecommunications Act 1981.

[4] Telecommunications Act of 1984, Chapter 23, Part V, § 69, "Dissolution of British Telecommunications."

[5] Department of Trade and Industry, *Communications Liberalisation in the UK; Key Elements, History* (March 2001), available at http://www.dti.gov.uk/cii.docs/com.lib.pdf.

[6] EU Com(87), 290 Celix No. 587 DC0290, "Toward a Dynamic European Economy," Green Paper on the Development of the Common Market for Telecommunications Services and Equipment.

[7] *Telegeography*, p. 9 (Telegeography 2000).

[8] *Telegeography*, p. 10.

[9] http://www.wto.org.

[10] Ley Organica de Telecomunicaciones (March 28, 2000).

[11] *Telegeography*, p. 123.

[12] Telecommunications Law 30/4/97 Government Decrees of 6/10/97 available at http://www.fedcomcom.ch.

[13] Constitutional Amendment No. 8/Aug/95. (The amendment relaxed the federal model that had previously existed under the Federal Constitution.) The Ministry of Communications added to this by presenting the PASTE restructuring initiative (Program for the Recuperation and Expansion of the Telecommunications Systems) from 1995–2003. The PASTE program estimated that $91.0 billion would be spent on restructuring during that time period.

[14] Law No. 9,295 (July 19, 1996). The law is also referred to as the "Minimum" or "Specific" Law.

[15] General Law of Telecommunications, Law No. 9,472 (July 16, 1997).

[16] Dunning, Thad, et al., *Brazil Internet Development for Whom; Brazil Telecommunications Infrastructure*, Stanford, CA: Stanford University, 2000. (The shareholders in each of the 12 new companies received the same common and preferred shares as they had owned in the parent company.)

Selected bibliography

Cantelon, Philip, *The History of MCI, The Early Years, 1968–1986*, Washington, D.C.: Heritage Press, 1993.

Frieden, Rob, *Managing Internet-Driven Change in International Telecommunications*, Norwood, MA: Artech House, 2001.

Gartner Group, *Liberalization Status Among Telecommunications Markets Worldwide*, Washington, D.C.: Gartner Group, 2000.

Intven, Hank, *Telecommunications Regulation Handbook*, Washington, D.C.: Infodev, 2000.

3

The historical evolution of global telecommunications markets

This chapter builds substantially on the overview of global market access and the basic regulatory formula for the creation of a competitive telecommunications market that was discussed in Chapters 1 and 2. This chapter charts the historical evolution of global telecommunications markets in general to competitive telecommunications markets and then focuses on several specific case studies of closed, partially open, and fully open telecommunications markets.

The increasing openness of the world's telecommunications markets is likely to be one of the single most influential factors in the growth of the world economy in the next 5 to 10 years. In this regard, it is likely that more and more markets will continue to open to competition in accordance with the basic regulatory formula. As addressed throughout this book, this path makes it imperative for governments, service providers, and consumers to adapt to this changing environment now, if they have not already done so. In order to do so, a firm understanding of the historical underpinnings of the evolution that has occurred and is ongoing in global telecommunications markets is required.

In order to assist in this effort, this chapter provides a brief historical perspective on how the majority of global telecommunications markets were initially structured. In addition, it focuses on the incremental steps that have been taken by many governments to create effective competition in telecommunications markets, including important considerations that have impacted this development. As discussed below, history demonstrates that in almost every country where there has been only what can be termed as "naturally progressive" improvement or growth in that telecommunications market, or in the rare case where the telecommunications market has seen a decline, the cause has been the failure of these governments to evolve their markets to competition.

The latter part of this chapter, in order to provide some real-life examples of the historical evolution of telecommunications markets, contains several case studies where this path has led to or is in the process of leading to the successful creation of competitive telecommunications markets, including that of the United States, Singapore, Australia, and Germany. This chapter also includes case studies of telecommunications markets, such as Vietnam and Jordan, that have remained primarily closed to competition or have only recently begun to open their markets to competition.

3.1 The evolution of telecommunications markets

3.1.1 The beginning

At the very dawn of the birth of telecommunications, in most countries, any entity that wanted to establish a telephone network could essentially build and operate its own. In some cases this would require the network owner to obtain a government franchise or other authorization. In many other markets, however, all that was required was a simple form of formal authorization. Because of the different standards, wiring, and network equipment being utilized by the multitude of players, quite often these networks were technically unique, so that consumers who utilized one network would find it impossible to access a user on a competing network that was not technologically compatible with the other network. Further, even in cases where technologically no issue existed regarding interconnection, some networks still did not interconnect because of the inability of the network providers to agree on commercial terms or simply the lack of desire of the operator to interconnect with additional networks.

Of course, this lack of ability to communicate across networks led to frustration on the part of consumers. Governments (including government users), businesses, and residential users soon began to recognize that a unified network was required to ensure seamless and reliable communications required for public safety, industry uses, or just plain conversation.

Accordingly, it soon became apparent to governments that the ability for people to communicate with one another over the telephone was dependent upon the use of either a single network, which provided nationwide coverage, or a group of regional or local networks, which were interconnected and fully interoperable. However, the former approach was favored and ultimately pursued. This was because of the overall government view of most countries that telecommunications services were essential services to the public (much like electricity or other utilities) and because of the complexities of infrastructure build-out and interoperability issues. Accordingly, most governments translated the creation of a unified telephone network into a decision to create a "natural monopoly" for the provision of telecommunications services.

The simultaneous decision that had to be made was whether this entity should be government or privately owned. This decision turned, in part, in most countries, on the recognition that this monopoly required some form of regulatory incentives and safeguards or the monopoly provider would not assure the universal availability of reliable, essential services at reasonable rates to the overall population. Accordingly, almost all governments believed that in order to meet the requirements of obtaining reliable telephone service at reasonable rates, a government-owned and government-operated service provider was required, because a commercial entity might not be willing to meet these mandates.

Governments also believed that it would be most efficient to combine telecommunications services with postal and telegraph functions and create a single government-owned entity to provide all three types of services. Governments believed that this grouping was appropriate, because all three functions were communications related. These entities became known as postal, telephone, and telegraph entities (PTTs) and, in regard to telecommunications, combined both the service provider functions and those of government oversight. Only a few markets, such as the United States, did not follow this model in the provision of telecommunications services by a government entity. Accordingly, during the early twentieth century, the private provision of telecommunications services was all but eliminated, and those functions began to reside with the government.

The PTT model, initially, was somewhat successful. Telecommunications services were generally limited to voice communications, which normally required a relatively simple telephone network, which the PTTs were able to build out fairly quickly to the main population centers. Most urban areas were able to receive basic telephone service, with high-level government offices, business customers, and higher-income residences being the primary customers that service reached. Quite often the lower-income urban residences and rural area residences were left underserved either because of the actual or perceived inability of customers in these areas to pay for service or the lack of network reach. Concern for this lack of service was rarely a government focus, because lower-income consumers were not always the most vocal or politically powerful, so their cries were often unheard or largely ignored by the political forces.

Nonetheless, in the mid-twentieth century, some countries began to attempt to rectify the problem of limited access to telecommunications services through the institution of early universal service programs. Unfortunately, many of these programs were unsuccessful in achieving even near universal service, since these customers were often the least financially attractive to serve, because of the cost involved in building the infrastructure to serve them and their questionable ability to pay for services. Although there were exceptions, such as in the United States, those exceptions were almost exclusively limited to higher-income countries. Accordingly, the bulk of service provision by the PTTs remained in urban areas, to high-income consumers, during the better part of the twentieth century.

As the twentieth century progressed, it became apparent that the monopoly service providers were not even able to adequately meet the service requirements of their urban customers and other high-end customers. Because of the lethargy of the monopoly service provider, network build-out was generally very slow. In addition, for example, because of the noncommercial nature of the operators, the PTTs often took years to fulfill orders for new telephone lines to be installed in residences or businesses. For example, in Buenos Aires, Argentina, prior to the privatization of the monopoly operator, it would take several years to receive a single residential narrowband phone line, even in a portion of the city that was already "wired." In Buenos Aires and similar markets, a black market in "wired" apartments was created, with large premiums being paid for apartments with phones. For the poor, who could not afford such premiums, or others, who refused to pay, theft of service was often the only alternative method to achieve telephone service. Accordingly, the Buenos Aires of even 20 years ago was a city of unplanned, unseemly wires that randomly crisscrossed the city as people hacked into the few existing phone lines. Unfortunately, in some closed markets today, those wire mazes still exist.

In addition, pricing for service was largely politically based, making for inefficient operators. Often, the incumbent service provider charged almost nothing for local or domestic calls (irrespective of cost), subsidizing these services with the funds raised for the much higher than cost priced international calls. This pricing strategy most negatively affected the international business community, which found itself held captive by the outrageous rates for generally poor services that the incumbent service providers charged. However, it did provide good fodder for political campaigns of candidates who could remind the citizens of the low rates for domestic phone services.

Further, often the PTT infrastructure was limited in reach and quality. Quite simply, in many countries, incumbent service providers did not have the funds or the political will to expend the funds to build out their network to serve all the customers who desired service or provide them with adequate service. Accordingly, often the infrastructure was limited to such an extent that multiple customers had to share party lines, or they could not get an open circuit when needed to complete an outgoing or receive an incoming call. In other cases, incumbent service providers were provided with no incentives to finance necessary upgrades to the network to ensure the telecommunications backbone worked effectively in order to supply more advanced services to customers.

In many of the developing world markets, such as Zimbabwe and Zambia, even today, party lines and busy signals are quite often the norm. This is, however, starting to improve, as many customers for voice telecommunications services are able to utilize less infrastructure-dependent mobile wireless services. Unfortunately, as discussed later, this has not greatly assisted those in need of the increased bandwidth requirements that many of today's customers require, especially in areas where even wireless infrastructure build-out has yet to appear.

The PTT as sole service provider model was the norm in most parts of the world until the 1980s. As discussed next, it was only with the dawn of the computer age, the improved ability to access the spectrum for wireless communications, and the increasing globalization of large telecommunications users and telecommunications service providers that the PTT model began to slowly fade away.

3.1.2 The dawn of the computer and wireless age: increased globalization means pressure on the traditional PTT model

In the latter half of the twentieth century, several major pressures for change began to be exerted on the traditional PTT model. First, the computer age led businesses to rely increasingly on data communications to run their businesses. This required improved infrastructure to carry increasingly broadband communications. In addition, the increasing advanced state of technology provided cheaper infrastructure for service providers, especially in terms of switching equipment. Third, the increasing ability to use wireless technologies for voice communications and more recently broadband for data services provided potential PTT competitors with a cheap and easy method to build the telecommunications infrastructure.

Another key change was the increasing late globalization of the world. This change really began to take effect in the 1950s, when air travel was no longer the province of the privileged few. On the contrary, at this time, many business people relied extensively on the ability of air travel for business meetings and to reach increasingly global subsidiaries and affiliates. Air travel made the world a dramatically smaller place to do business in. Second, as mass media improved, places that seemed foreboding were opening up in people's living rooms. Further, as the Cold War began to wane, new markets became available for companies to both sell and manufacture their wares. All of this meant that the world was shrinking from a business perspective, and industry found it increasingly possible and desirable to operate on a global basis. However, such operations required improved communications between global sites, something that was not possible under the traditional PTT model.

In addition, with the advent of the computer, improved wireless technologies, and the increasing globalization of the business world, pressure was placed on many governments (primarily in developed countries) to begin to consider ways to improve the quality, reach, and pricing of their telecommunications networks. This reevaluation was caused by many factors. First, businesses began to voice their concerns about operating in a country where obtaining a phone line may take months or, in extreme cases, even years to be installed. Other business customers were also concerned about the reliability of the telecommunications infrastructure and the ability to obtain new services, such as broadband data services. For example, in many markets, where the business climate was fairly well developed, businesses were still hesitant to enter, because of a lack of access to more sophisticated telecommunications services and the unreliability of the services that existed. Also, because of the traditional methods of pricing telecommunications services, many businesses rightly believed that they were being required to subsidize a country's total communications requirements, simply because they were able to afford more services and had international telecommunications needs.

Another incentive for change was the recognition by entrepreneurs, existing service providers, and governments that the provision of telecommunications services is a very lucrative business. The service providers and entrepreneurs realized that large segments of the population, including high-end international business customers, were lacking basic and advanced services that would bring in large revenues, even if offered at lower rates than the incumbent service providers. This pricing scenario was possible because of the high profit margins the incumbents typically operated under and the efficiency of their operations. Both entrepreneurs and existing service providers recognized that many efficiencies were lacking from the current model that they could take advantage of and that would allow them to provide better services and vastly more reasonable prices.

In addition, the post-World War II years brought a recognition that open international trade was necessary for the economic survival of a country. Accordingly, governments began to liberalize the overall requirements of foreign companies to operate in their markets. Governments also recognized that by allowing competition into the telecommunications market, they could gain revenues through privatizations, licensing fees, taxes, and other fees. In addition, by opening telecommunications markets to competition, new higher-paying jobs would be created in the marketplace by both new service providers and the businesses that they would serve. Accordingly, many governments began to recognize that financially opening their telecommunications markets to some form of competition could be rewarding.

Further, the post-World War II world also encouraged increased political and economic relations with other countries. With the founding of the United Nations in 1945, the increasing efforts at forming alliances with both wartime allies and enemies, and the formation of free trading areas, such as the European Union, companies began to find it easier to have global operations and governments work together on areas of strategic economic importance. The end of the Cold War in the latter part of the twentieth century and the subsequent fall of Communism in the 1980s also led to increased globalization, through the opening of trade and improved political relations with formerly closed markets.

Following this market opening, to date, most governments that have introduced competition followed the basic regulatory formula discussed in Chapter 2. Many other governments are at various stages of this process. Other governments, as will be seen in the case studies, are still stuck at the first stage.

3.1.3 The political and legal decision to move forward

The first step in moving forward with creating effective competition in a telecommunications market is making the political decision to proceed. Many markets have failed at their efforts because of their inability to make this decision and the lack of will to implement it. An example of this is Costa Rica, which has been toying with the concept of liberalization and privatization since the early 1990s but, because of political considerations, has failed to take action. Action has been stymied in large part because of the politically influential labor unions that control the employees in the telecommunications market; these unions have threatened serious repercussions if the government moves forward with any form of market opening.

Many other markets, though, have been able to take tough political decisions and move forward with market opening. Once this decision has been made, some governments moved quickly forward with implementation, while others, such as Australia, go through extensive study periods leading up to the necessary changes. In either case, once the implementation process occurs, there may be many legal changes required for market opening to proceed.

In terms of legal structures, many countries have a constitution or law that awards exclusivity for telecommunications services to the incumbent service provider. For example, as discussed in Chapter 2, in Brazil, constitutional changes had to be made in order to allow the government to go forward to introduce any significant form of competition in the market. Similarly, once the constitution was amended, the Brazilian legislature had to pass whole new telecommunications laws to govern the newly forming telecommunications markets. Only after these types of changes are made can the regulator be established and move forward on creating a regulatory structure that works within the confines of the overarching laws. In many countries, such as Costa Rica, it has taken years for these types of changes to be made. For example, Costa Rica has been unable to pass a law since its initial tries in the 1990s that would even privatize its incumbent service provider, much less introduce competition in its telecommunications markets. However, other countries, such as Chile, were able to make these monumental changes in compressed time frames. Quite often, timing depends on the political will of the country for such change.

Similarly, complementary laws, such as those addressing the repatriation of profits and taxes on services, the ability for foreign investment, and the ability to establish a local entity, may also need to be amended in order to make investment opportunities more attractive. A case in point is India, where current import taxes are upwards of 100% on the importation of telecommunications equipment. Such laws act as barriers to legal entities and, as discussed later, must be addressed in order to have true competition.

3.1.4 The initial incremental step: establishing and implementing a regulatory regime

As the overall legal structure is created for opening the telecommunications markets, most governments start with the same first step: separating the service provider and regulatory functions. This is an important step, since it ensures that an independent regulator will regulate the incumbent service provider through the privatization and liberalization stages. In addition, it is a mechanism to ensure that the overall policy and goals of the country are met through the regulatory process.

Past efforts at creating an independent regulator have demonstrated that how this step is implemented is of paramount importance to the success of further efforts to open the telecommunications market. First and foremost, the regulator must be isolated to the greatest extent possible from the political considerations of the government and be able to manage its budget independently. A good model of this is Anatel, the independent telecommunications regulator in Brazil. The opposite has been true in Mexico, where COFETEL has been a largely ineffective regulator because of its traditionally close ties, which it has not severed, with Telmex and because of the political nature of the regulatory process, where Telmex has been able to garner strong and powerful political allies.

In addition, the regulator must be given an adequate budget for staff and other resources and the staff must be qualified to perform its work. Most countries have limited personnel, outside of the incumbent service provider, who have expertise in telecommunications. Accordingly, often the regulatory body is initially staffed by former members of the incumbent service provider or by staff seeking employment in the telecommunications industry in the future. Ethical rules are often enacted to try to curb any undue favoritism toward the incumbent service provider from the staff. For example, in the United States, government regulations provide limits on matters that ex-FCC staff may work on once they leave the FCC's service. Often, regulators hire staff from other disciplines, whose past expertise may assist in the new regulatory efforts at hand. For example, disciplines such as energy regulation often have similar concepts, which are transferable to the telecommunications regulatory arena.

Past experience also demonstrates that the establishment of a well-functioning regulator takes time. In many countries, the ministry of telecommunications or other executive-level entity either works with the regulator or provides certain regulatory functions. For example, in Japan, the Ministry of Posts and Telecommunications has responsibility for overall telecommunications policy and guidance, while the Telecommunications Technology Committee determines the communications protocols and interconnection standards. Often, as the regulator matures, the importance of the telecommunications ministry diminishes on at least a day-to-day basis. In addition, once established, a regulator must create a day-to-day bureaucracy, draft rules, implement policies, and explain decisions. These rules may take months to establish and to get operating.

3.1.5 The next incremental step: privatization of the incumbent service provider

Historically, the next incremental step in the evolution of the telecommunications market has been that of privatizing the incumbent service provider. An evaluation of global markets where this has occurred reveals that this step has also been generally successful. In 2000, for the first time, the number of privately owned incumbent service providers outnumbered the number of government-owned service providers [1]. As discussed in Chapter 2, in most countries where the privatization process has taken place, these markets have seen improved service, better infrastructure, and slightly lower prices. As will be discussed throughout this book, markets that have liberalized but have failed to privatize their incumbent service providers often face issues of anticompetitive conduct by the incumbent service provider, because of the inherent advantages that a government-run service provider has over private competitors.

ing the dominant abuses of Telefonica de Peru remain, the Peruvian market has become increasingly competitive.[1]

The liberalization step has traditionally required a government to determine what services to make competitive and whether it wants to encourage facilities-based versus non-facilities-based competition. Other key decisions involve the number of entrants it wishes to allow in for each type of service provision and the timing of such entry. Historically, most countries are likely to allow full competition in the value-added services market but are substantially more likely to impose limits on the number of basic telecommunications service providers, including mobile wireless operators.

Further, to ensure that new entrants are able to compete against the incumbent service provider, history has demonstrated that it is essential that the government institute competitive safeguards. Key safeguards include instituting interconnection guidelines, mandating number portability, and providing for equal access and presubscription, among other items. In markets where such safeguards have not been instituted or have not been implemented in a meaningful manner, such as in New Zealand, even where competition was fully allowed, new entrants have often been stymied in their efforts to capture any meaningful amount of market share. However, in markets where the government has committed to fully institute competitive safeguards, such as in the United Kingdom, competition has been able to blossom.

3.1.7 The continuing step of creating effective regulation

Throughout this entire process, there has historically been a need to continue to refine and revise the regulatory regime to take into account the market evolution. Accordingly, the regulatory regime that was appropriate at the start of competition may not be appropriate as competition evolves. For example, the United States initially required non-facilities-based service providers (resellers) to obtain affirmative authorizations to provide domestic serv[illegible]s. However, because of the increasing competition in the market [illegible] was able to loosen this requirement for nondominant

[illegible]rovider of telecommunications in Peru. It provides fixed local, [illegible]nal long-distance telephone services, as well as cable TV and [illegible]ruvian market is increasingly a competitive market. Specifi[illegible]iness has produced lower rates. "Fitch Downgrades Telefon[illegible]cy to BB–." Updated June 5, 2001.

In some cases, however, rules may need to be revisited throughout the liberalization process in order to make them stronger, or new rules are required to prevent anticompetitive abuses. For example, in New Zealand, the government revisited its initial decision to minimally regulate its telecommunications market. In 2000, the government announced that it needed to revisit this format because of the lack of competition that had been created in the telecommunications market. Since this announcement, New Zealand has begun a series of regulatory revisions to ensure that safeguards are in place that foster the ability for new entrants to compete against Telecommunications New Zealand Ltd. [2].

In very few cases has deregulation truly been seen in the regulatory regimes of countries seeking to reach an effectively competitive market. This is because in the early stages of competition, increased regulatory scrutiny is generally required to ensure that a level playing field is created. There have been some limited exceptions, such as New Zealand. The government of New Zealand took a very novel approach by fully opening its market to competition, without creating a regulatory agency [1]. Instead, the only curb on the market was the Antitrust Commission. Unfortunately, this approach has proven unsuccessful, since Telecom New Zealand was able to act in an anticompetitive market without a direct regulatory check. This resulted in a market where only a small percentage of customers changed service providers. As discussed above, these failures led the New Zealand government to recently institute a new, stronger regulatory regime to govern its telecommunications market.

In addition, regulation must adapt to the changing industry. The increasing importance of telecommunications in the world, with changes brought by technological innovation and market expansion in other communications-related sectors, is leading to increased convergence among the communications sectors. Accordingly, new regulatory devices are necessary to handle this trend. In some countries, this has already begun to happen. For example, in the United Kingdom, the government announced that it would eliminate its current regulator, OFTEL, and replace it with a modern regulator more attuned with the technological advances that have been made since OFTEL's creation [3].

3.2 Examples of market evolution

In this section, the focus is on the historical evolution of both successful and near successful market opening in all regions of the world, as well as markets that have failed to open to competition. As will be demonstrated, each market must be evaluated in its own cultural, political, and legal context.

3.2.1 The Americas

The Americas have made substantial inroads in creating competitive telecommunications markets, especially in the past couple of years, where the opening of key markets has included Canada, Argentina, Venezuela, and Brazil. For example, ITU data show that the Americas have the largest percentage of privatized incumbent service providers in the world, with 63% in 2000 [4]. Further, the Americas remain the most open to foreign investment. The following text focuses on four examples: the United States, Venezuela, Honduras, and El Salvador.

The United States

Although discussed briefly in Chapter 2, the telecommunications market evolution in the United States is expanded on in this section because of its importance in setting many of the standards used around the world in establishing effective competition in the telecommunications market. The United States has played a leading role in seeking to open global teleommunications markets to effective competition.

The government of the United States was one of the first to create an effective competitive telecommunications market. However, this market evolution occurred in a slow phased approach, with a hesitant regulator, very aggressive new service providers, and a tremendous push from the U.S. judicial system. In addition, the U.S. telecommunications market was traditionally unique in that it had an established regulator from the early twentieth century and had a privately owned incumbent service provider.

In its early stage, the U.S. telecommunications market was, by appearance, open to competition, with numerous independent service providers and one large operator, AT&T and its affiliates, operating around the country. Often, these service providers could not interconnect with one another. This inability often led to frustrations on the part of customers. Customers of AT&T, however, did not face these same problems when talking to another AT&T customer, even across the country, because of its affiliation with local service providers.

Further, AT&T had control of many essential patents on voice transmission technology and patents on long-distance communications technologies. Through the control of these technologies, which meant that

AT&T was able to provide improved service, AT&T and the Bell System were able to establish themselves as the dominant providers of telecommunications services in the United States. By 1907, Bell companies were operating almost half of the telephone businesses in local markets.

Early on, AT&T forced independent phone companies to either fold or become a part of its Bell system by providing superior long-distance service to its affiliates but not its competitors. Although the U.S. government never granted AT&T an exclusive franchise for the provision of long-distance services, through its anticompetitive conduct it was able to gain a majority of the market share in this sector by these acquisitions and its consistently better service. In addition, through its affiliated Bell operating companies (BOCs), AT&T provided local service and access to long-distance service in a large portion of the United States. Although a number of independent telephone companies provided local telephone service and access in certain locations, they were reliant on AT&T for long-distance connection to other local companies. AT&T also controlled most of the manufacturing of telecommunications equipment through its affiliate, Western Electric, and was a major force in telecommunications research and development through its research arm, Bell Laboratories. This control and influence further allowed it to tighten its grip of control over the U.S. telecommunications market.

In the late 1920s, it became apparent that AT&T was a force to be reckoned with, since it truly controlled the lion's share of the U.S. telecommunications market. However, because of U.S. hesitancy toward government ownership of businesses, AT&T continued to be a private service provider. However, in order to help curb the potential anticompetitive impulses of AT&T and to ensure that the communications network met the public policy goals of the United Sates, the Communications Act of 1934 was established. The U.S. government recognized, during the time of the New Deal, when there was increased scrutiny of monopolies, that AT&T's dominance left the door wide open for it to abuse its dominant position to the detriment of consumers.

The Communications Act of 1934 accomplished several major goals. First, it created the FCC, which could oversee the conduct of AT&T [5]. Second, it established a mechanism in which to control the rates that AT&T was allowed to charge its customers [6]. Of equal importance, it established the concept of universal service, which ultimately provided the United States with the highest teledensity in the world [7].

However, the Communications Act of 1934 failed to explicitly mandate competition. Accordingly, as discussed below, when market and technological changes finally permitted the first attempts at introducing competition into the U.S. telecommunications market, they were met with a hesitant regulator and an aggressive AT&T.

Between the passage of the Communications Act of 1934 and the early 1980s, a series of developments in the United States led to the introduction of effective competition in much of the telecommunications market, including the provision of national and international services. First, technological progress (such as the development of microwave systems and computer technology) opened ways for new entrants to enter the telecommunications market and allowed these providers to offer competing services at lower prices than AT&T. For example, wireless infrastructure began to become a realistic technology in which to carry commercial telecommunications traffic. Because of the need for substantially less and cheaper infrastructure (i.e., no terrestrial wires to install), the services were less costly to provide. Second, consumers were beginning to demand more from the telecommunications market and AT&T, in terms of new services, lower prices, improved customer service, and the like. AT&T, because of its lethargy from its monopoly status, was slow to respond. However, other companies, watching from the sidelines, began to see this as a market with unlimited opportunity.

The FCC also recognized the need to adapt to new market demands, albeit very slowly. Beginning in 1959, the FCC began allowing private companies to use microwave technology to carry their own long-distance traffic (i.e., private networks). Accordingly, a company could actually set up its own intracorporate communications network, using microwave technology, and bypass the public switched network. In 1964, Microwave Communications Inc. (MCI) applied to the FCC for authorization to construct its own network to carry private companies' long-distance traffic. MCI recognized that there were certain synergies inherent in its ability to commingle traffic of multiple companies, including the ability to offer private network services at dramatically lower prices than AT&T. Subsequently, MCI sought FCC authorization to carry all types of long-distance calls and to interconnect with AT&T's network. The FCC sought to limit these activities through a series of decisions, forcing MCI to take its case to the federal courts. The courts sided with MCI, hence opening the doors to competition even further.

A pivotal point in the development of competition in the U.S. long-distance market was the breakup of the Bell system as the result of a consent decree, which terminated a U.S. government antitrust suit against AT&T [8]. As discussed, AT&T owned the BOCs local networks, which provided a substantial number of the telephone lines in the United States. As a result, AT&T's long-distance competitors had to interconnect with their local networks to originate and terminate calls. In 1974, the U.S. Department of Justice filed an antitrust suit against AT&T to halt these anticompetitive practices. In 1982, after a protracted battle, the U.S. government and AT&T negotiated a consent decree, which became known as the Modification of Final Judgment (MFJ) to settle the suit. The MFJ required AT&T to divest the BOCs and placed certain line of business and other restrictions on the BOCs, such as not being permitted to manufacture telecommunications equipment and customer premises equipment (CPE) and the requirement to provide long-distance competitors equal access to the local network (i.e., access that was comparable to what it was providing AT&T in terms of terms and conditions). A significant part of the settlement was that AT&T was permitted to retain Western Electric and the Bell Laboratories. The FCC, in reaction to the consent decree, was required to revise its rules and policies significantly to ensure that long-distance service competition was a real possibility. Numerous proceedings were conducted concerning areas such as number portability, dominant carrier safeguards, the establishment of accounting principles, and others.

Because of these developments, competition was provided with a framework in which to fully blossom in the U.S. long-distance market. Nonetheless, competition was slow to begin. Because of the long-time relation AT&T has had with its customers, AT&T has been able to retain a dramatic portion of its market share—ensuring its continued dominance in the long-distance service marketplace—even to this day. The ability of an incumbent service provider to retain market share even with an effective regulatory regime in place is not a phenomenon that has been limited to the United States. On the contrary, in all but the rarest cases, competition is slow to erode the incumbent service provider's market share.

Competition in the local market, however, has taken longer. It was only in 1996 that the Telecommunications Act of 1996 (1996 Act) allowed competition in the U.S. local market to also develop. In the 1996 Act, the U.S. Congress acted to mandate that most significant economic impediments to efficient entry into the monopolized local market be removed by the FCC. Specifically, the 1996 Act mandated the use of three key methods of entry into the local market: the construction of new networks, the use of unbundled elements of an incumbent local exchange carrier's network, and resale. Since 1996, the opening up of the local market in the United States has not allowed in too much competition. The reasons that are generally cited for this result are the continuing grip of the former Bell operating companies through anticompetitive conduct and, more recently, the downturning of the U.S. economy. Nonetheless, it is likely that competition will increase in this market as time progresses and service providers have greater expertise in entering this market.

Venezuela

Venezuela is a key Latin American market. However, throughout the past couple of decades, it has been subject to a series of political and economic crises that have limited its growth. Traditionally, Venezuela has had a vibrant oil industry. Because of this industry, and the need for businesses that operate around it, telecommunications has always been an important market. However, because of political concerns, competition has been slow to evolve.

The Venezuelan telecommunications market was dominated by the state-owned incumbent telephone company, Campania Anonima Nacional de Telefonos de Venezuela (CANTV) until recently. In 1991 the government began to open its market, first by creating by presidential decree a regulator, Comision Nacional de Telecomunicaciones (CONATEL), under the jurisdiction of the Ministry of Transport and Communications. CONATEL is comprised of seven members and one director, who is appointed by the Ministry of Infrastructure. CONATEL is responsible for spectrum management, granting of concessions and permits, bilateral and multilateral matters (such as relations with foreign long-distance carriers, satellite operators, and international agreements), and establishment of technical standards for equipment. That same year, the government took a first step in partially privatizing CANTV, completing the privatization process in 1996. CANTV was awarded a monopoly for basic services until November 2000. However, this was somewhat modified in 1997 when the government began to allow limited competition for the provision of rural telephony services and before that in 1991 when the market for value-added services was opened.

With its eye on the 2000 introduction of competition, CANTV made substantial network investments. For example, between 1991 and 1997,

CANTV added 1.8 million fixed lines, to bring the total capacity of its network to 3.41 million households. This represented a teledensity of 11.77 lines per 100 inhabitants. CANTV also spent substantial funds on digitizing its network [9].

CANTV also prepared for competition by reducing the levels of its international tariffs and domestic tariffs. Along with this process, CANTV introduced a tariff rebalancing process to bring its long-distance and international service tariffs more in line with actual cost.

The Venezuelan government has lived up to its promises, including its WTO commitment, and in 2000 the market for basic services was fully opened. In order for this to occur, a new telecommunications law was enacted on June 12, 2000 (the old law was from the 1940s and had not been substantially updated). This law was expansive and provided for competition in all areas of telecommunications, including the framework for competition in the basic service market. To date, various companies have received authorizations from CONATEL, including New Global Telecom, Convergence Communications, Veninfotel Comunicaciones, and Tercel C.A. With competition just beginning, it is hard to evaluate the success of the market to date.

Honduras

A less-successful example of liberalization and market opening is Honduras. This in part can be attributed to the less-developed nature of the Honduran economy. The per capita GDP is substantially lower than the more-developed Latin American economies of Brazil and Argentina. Teledensity has also been historically low, with only 4.42 lines per 100 inhabitants in 2000 [10]. In addition, Hurricane Mitch in 1998 caused substantial damage to Honduras's telecommunications network.

Empressa Honduerena de Telecomunicaciones (HONDUTEL) remains the government monopoly for basic communications services, with past privatization efforts a failure. Nonetheless, the government announced plans to have a new offering in 2001.[2] However, the value-added service market has been open to competition.

Honduras has two government bodies with jurisdiction over telecommunications. The Ministry of Communications, Public Works, and Trans-

---

2. In a May 20, 1999, debate, the Honduran Congress approved amendments to the HONDUTEL privatization authorization law. These amendments authorized a stock offering of 51% to an international operator.

port is the policy maker, and Honduras has also established a day-to-day regulator, the Comision Nacional de Telecomunicaciones (CONATEL).

Foreign ownership, because of Honduras's complicated legal structure, remains restrictive. Currently, foreign companies are required to act through a local agent, which is 51% locally owned, and foreign investors are taxed 5% higher than locals. In addition, special government authorization is required for investment.

El Salvador

Although El Salvador is a lesser-developed country, it has been an early Central American market opener to competition. After several years of civil war and devastation from a large earthquake in the mid-1980s, El Salvador's telecommunications sector is thriving as the country has begun to have economic growth. Beginning in 1996, with the passage of the new telecommunications legislation, the government of El Salvador embarked on a path of liberalization that culminated in the July 1998 sale of the state-owned telecommunications company—ANTEL [11]. ANTEL was divided into two separate companies—Compania de Telecomunicaciones de El Salvador (a wireline company plus a Band B PCS license) and Internacional de Telecomunicaciones (INTEL—a wireless company with a cellular B band concession). CTE and INTEL were sold to France Telecom and Telefonica de España, respectively [12].

Since this time, El Salvador has also allowed competition in all sectors of the market. Competitive entrants in the market include Telefonica El Salvador, Vianet, Americatel, Amnet, GCA Telecom, Telpro, Saltel, Salnet, Telecam, and Publitel. This market opening was embodied in El Salvador's WTO commitment, which provided for full competition for all basic telecommunications services for all market segments. El Salvador also fully committed to the WTO Reference Paper.

The primary authority for telecommunications in El Salvador is the Superintedencia General de Electridad y Telecomunicaciones (SIGET). SIGET's principal duty is to enforce compliance with laws and regulation, ensure universal service, issue licenses, and manage the spectrum resource, among others. SIGET is a completely autonomous institution of the government, but the Minsterio de Economia represents SIGET in the National Assembly and pays its annual budget [13].

It is still unclear whether El Salvador will live up to its promise through the institution of effective regulations to curb potential anticompetitive conduct by the incumbent service providers. The recent earthquakes have created some barriers to competition, because of the disturbance of the physical telecommunications network. Nonetheless, it appears that competition is beginning to take hold. For example, at the end of 1998, there were 380,000 wired lines installed in the country and 70,000 cellular phones in operation [12].

3.2.2 A new, unified Europe

Perhaps one of the greatest pressures for global telecommunications markets to liberalize has been the overall market opening of the European Union and the 1998 requirement of the introduction of competition in the EU member states telecommunications markets. (See Figure 3.1 and Table 3.1.) A comprehensive evaluation of the countries to date reveals that both incumbent service providers and new entrants have met with success.

In terms of the incumbent service providers, most took advantage of the time between the announcement of the introduction of competition and the actual introduction of competition to change from a bloated bureaucracy to more streamlined commercial operators. Many of these operators, while seeing their overall market share decrease, have seen increased revenues from streamlining their operations, becoming more efficient operators, and introducing new, more attractive services.

In terms of new entrants, while quite a few have had rough starts, many others have been able to well position themselves in the new market

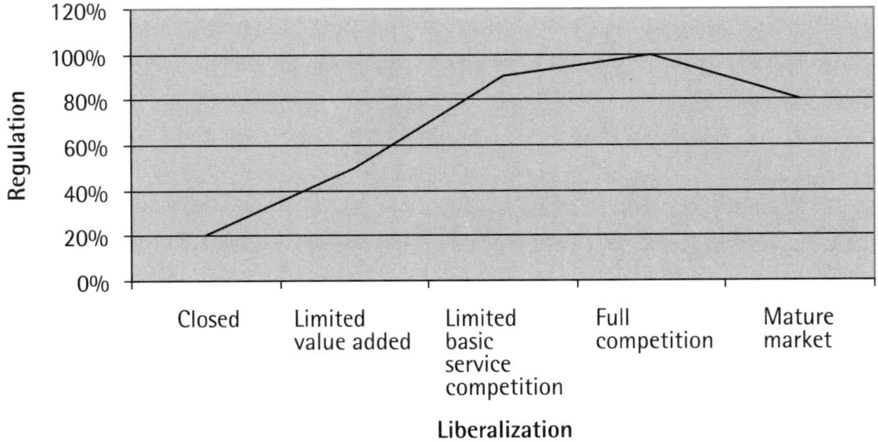

Figure 3.1 Liberalization chart.

TABLE 3.1 EU Liberalization: January 1, 1998

Ended right of monopoly for voice telephony
Required the establishment of independent regulatory authorities
Required transparency of tariffs and interconnection terms
European Commission has taken steps to ensure that liberalization is on track
The United Kingdom and Finland liberalized prior to the January 1, 1998, date; several countries liberalized after the date, including Greece and Portugal.

order. For example, U.S.-based WorldCom, Inc., has opened facilities-based operations in almost every major European market. Other service providers, such as Global Crossing, have made similar inroads. In addition, many of the incumbent service providers have found themselves well positioned to attack the other EU member state markets. This has been particularly true of France Telecom, through its ownership of Global One, and British Telecommunications, through its participation in Concert.

In this section, the German market is reviewed first because of its particular historical importance, since it involved the partial privatization of a major force in the German market, Deutsche Telekom (DT), with some continued government ownership, a new regulator, and the battle for a very lucrative market by new entrants.

These same battles have been repeated in many EU markets, as well as the markets that are currently seeking admission to the EU, such as Hungary and the Czech Republic. As will be discussed further, because of the problems that have arisen in markets such as Germany, the European Union is currently reviewing the need to institute a major revision of telecommunications regulation. Such an approach could include transferring the generic regulatory authority for member states to a European Union–level body.

This section also discusses Hungary as a case study. Hungary is of particular interest because it is currently an associate member of the European Union. As such, it is required to bring its legal and regulatory structure in line with that of the European Union, including the area of telecommunications.

Germany: a slow opener

Germany is one of the richest countries in the world, with an estimated GDP of $1.864 trillion and GDP per capita of $22,700 in 1999 [14]. Germany is a powerhouse in the European Union telecommunications market, with approximately 25% of the market share [15].

The German telecommunications market was traditionally dominated by the incumbent service provider, Deutsche Bundespost. In response to increased pressures to liberalize its telecommunications market, the German government separated the Bundespost into three entities in 1990, including Deutsche Telekom (the other entities address the postal and telegraph markets) [16].

Deutsche Telekom was partially privatized in 1996, with a second round in 1999. However, following even the 1999 round, the German government still controlled approximately 65% (direct and indirect) of the company. Today, Deutsche Telekom has a percentage of continuing government ownership. The continuing presence of government ownership in Deutsche Telekom has provided competitors with increased support for their allegations of a biased regulator toward the incumbent service provider.

On January 1, 1998, the new German regulator was created, the Regulierungsbehorde für Telekommunikation und Post. Its functions include promoting and ensuring fair competition, preventing discrimination, encouraging market growth, and adequate communications services throughout Germany. It is also responsible for awarding authorizations and establishing interconnection fees [17].

In anticipation of meeting the January 1, 1998, date for EU telecommunications liberalization, the government enacted into law the Telecommunications Act on August 1, 1996. The Telecommunications Act focused on three items: number portability, preselection, and the provision of new operators to use Deutsche Telekom's local networks and equipment.

As is typical of incumbent service providers, prior to the introduction of competition, Deutsche Telekom spent substantial funds preparing for competition. By 1997, Deutsche Telekom had a 100% digitized network. In addition, it had achieved a teledensity of 55.6% [15]. In the early days of competition, Deutsche Telekom expended substantial funds on improving its long-distance network, the area that it was most vulnerable to from competition. In addition, in March 1998, Deutsche Telekom made large rate cuts to keep itself competitive. It also began to focus on the introduction of new services, such as Internet telephony.

To date, numerous facilities-based authorizations and infrastructure authorizations have been awarded with little regulatory burden. The most noted of these competitors includes Mannesmann Arcor and Viag Interkom. The German market for value-added services had been open since 1996. Numerous value-added service registrations have also taken place.

The initiation of competition was quite complex in Germany, however, because Deutsche Telekom established numerous roadblocks to the creation of a competitive telecommunications market. One of the first big issues to be resolved was the price for local leased lines obtained from Deutsche Telekom. Deutsche Telekom initially set a fee that the regulator found was excessive. When asked to reset its rate, Deutsche Telekom once again set an exorbitant fee. However, as opposed to overturning this fee while it examined what a reasonable fee would be, the regulator let it stand during the investigation process. This effectively meant that new entrants were unable during the investigation period to offer a competitive product until after this issue was settled [18].

Interconnection fees posed similar problems. Deutsche Telekom once again set extremely high fees, only to be overturned by the regulator. However, the fees ultimately established by the regulator were still deemed too high by new entrants and, not surprisingly, too low by Deutsche Telekom. Over time, this issue was resolved; however, it highlights the very real issues that new entrants have had to face to obtain a level playing field for telecommunications services in Germany [19].

Despite these problems, inroads have been made. Today, most service providers have been able to successfully build out facilities to provide service. Similarly, the problems associated with interconnection to the Deutsche Telekom network have abated. However, issues still arise that must be resolved at the regulator level.

Hungary

Hungary's financial health has been on the rise recently. With a population nearing 11 million people, it has a per capita GDP of approximately $4,500 [20]. In the mid-1990s, Hungary had a teledensity of 26.06 people per 100 main lines [21]. By 2000, it had grown to 40.24 [21].

In the early 1990s, Hungary began to restructure its telecommunications market. Prior to 1989, the state-owned Post, Telegraph, and Telephone (PTT) was responsible for all provision of telecommunications, broadcasting, and postal services in Hungary. In 1989, the government split the PTT into three companies, including MATAV,[3] which was to be

3. The other two entities included the Hungarian Broadcasting Company and the Hungarian Post Office.

the monopoly provider of telecommunications services. The first round of privatization of MATAV, the Hungarian incumbent service provider, occurred in 1993, with just over 30% of the shares sold to a consortium consisting of Ameritech and Deutsche Telekom [22]. In 1995, the consortium further increased its stake, with a small public offering also occurring [23]. The government still holds a small stake in MATAV. MATAV had a monopoly for domestic and international long-distance service until 2002. There is limited competition in regional or local calling, with MATAV still providing the bulk of local services. The Hungarian government has fully liberalized the value-added services market and competition in this market is vibrant.

The Communications Authority is the licensing and regulatory body for telecommunications, postal services, and frequency management in Hungary and operates under the Ministry of Transport, Communications, and Water Management. The Ministry is responsible for policy and crafting regulations governing the issuance of regulatory authorizations.

3.2.3 Asia Pacific

The Asia Pacific region has had an interesting but checkered history in introducing competition into its telecommunications markets. As discussed throughout this book, many of the most liberalized markets in the world, such as New Zealand, Singapore, and Australia, are in Asia. However, other markets, such as Vietnam and Thailand, have been resistant to allowing the introduction of competition. ITU data provide that in 2000, two-thirds of the region was still dominated by monopoly service providers. In the next sections, we examine the dichotomy of Australia, a country that made long-term commitments to liberalization, and the early decision of Singapore to open its market to competition as compared with Vietnam, which remains one of the most closed markets to private investment and competition in the telecommunications sector.

Australia

Unlike the United States, reform in the Australian market did not arrive until the late 1990s and the government, not the private sector, championed its cause. It is comparative to the United States in terms of territory—both tremendously large market areas. However, Australia is substantially more rural than the bulk of the United States. It has a strong economy based largely on earnings from commodities. In 2000, Australia's teledensity was estimated at 52.12 per 100 lines [21].

Although telecommunications service in Australia was first provided by the Postmaster General, the Telecommunications Act of 1975 created the Australian Telecommunications Commission (ATC). The Act granted ATC the exclusive right to erect and maintain telecommunications infrastructure. This system worked relatively well until the 1980s forced a government review of the market. Pressures that brought about this change included the development of and the need for the introduction of new services, the initiation of government reforms in numerous sectors, and the launch of the Australian satellite system AUSSAT, which created limited competition to the Australian monopoly service provider. Further, the WTO Agreement was in the early stages of negotiations and international pressure for change in the Australian telecommunications market was increasing.

In response, in 1988, the government issued a telecommunications policy statement on reform, which led the government to undergo a firm review of its telecommunications market, including the extent to which competition should be allowed. The government, however, did not feel comfortable with the United States or the United Kingdom's approach to competition. Specifically, the Australian government felt that the tariff rebalancing that had increased basic service rates for residential services was inconsistent with the policy objective of sustaining universal affordable access to basic telephone services. Accordingly, Australia had the difficult task of determining a path to competition, consistent with its own societal requirements.

In part because of this balancing requirement, change was slow to come. It was only in the mid-1990s that the Australian government created a new regulatory framework. This framework created an additional regulatory agency, called the Australian Competition and Consumer Commission (ACCC). The Australian Communications Authority regulates consumer and technical matters, manages radio communications, issues licenses, and regulates service providers. The ACCC acts as an arbiter and tries to ensure fair competition. The Australian telecommunications market is even more unique, because there are other government entities with responsibility over the telecommunications sector, including the Australian Communications Industry Forum, which is responsible for developing technical, operational, and consumer industry codes and standards, and the Telecommunications Industry Ombudsman, which is the independent arbiter for complaints made by residential and small-business consumers against their service providers. This is in comparison to most markets, where only one or two government agencies have a role in day-to-day telecommunications regulation.

The value-added service market opened in 1989. Although Telstra continues to hold the largest market share, competition has blossomed. Liberalization of the basic service market was introduced in 1991, when Optus was licensed to compete with Telstra. At the same time, the government authorized resellers and providers of private networks and corporate services to obtain domestic and international authorizations.

In 1997, the market was fully opened to competition, including facilities-based competition, slightly earlier than was publicized by the government. To date numerous service providers have been authorized to provide facilities-based services. However, these new competitors have had little impact on the ability of Telstra to continue to dominate the Australian telecommunications market. This ability to continue its dominance is based on several factors, including its incumbent position, its ability to upgrade its network substantially, the large network it has in place, and its entrenched customer base.

Optus, the first facilities-based competitor to Telstra, faced substantial regulatory hurdles. Telstra has been accused of poorly implementing number portability and access codes, which took several years to resolve. In this regard, Optus filed suit against Telstra under Article 46 of the Australian Trade Practices Act, which forbids a company to use its corporate power to eliminate competition [24]. Optus alleged that Telstra refused to supply local call access or local call resale on favorable terms, if at all, and of failing to implement local number portability on reasonable terms. It also alleged that Telstra was protecting its local call market by cross-subsidizing its broadband network and its pay TV service.

The first part of Optus's complaint was resolved in 1998, with an agreement with Telstra under which it could resell Telstra's local call services [25]. Simultaneously, ACA published a timetable for the introduction of nationwide number portability [26]. Other service providers also voiced their complaints. These suits included allegations of failure of Telstra to provide noncode access to a substantial number of customer line numbers. These complaints were resolved over time.

Accordingly, the development of competition in the Australian telecommunications market has faced many of the same hurdles as in other markets. However, the outlook for increased competition is promising, and the continued path toward the creation of a level playing field has been set. With many new competitors, Australia may be able to capitalize on its telecommunications market.

Singapore: a market that saw the light and changed paths quickly

Singapore is a small island nation with a population of approximately 3.2 million; however, it is an extremely developed country with a high-technology bent. Singapore's economy was not as hard hit as other Asian economies, in part because of its growing importance as a regional location for the headquarters of multinational corporations. However, because of this positioning, Singapore found that telecommunications costs, reliability, and availability became an increasing area of concern for these businesses. Singapore has a very high teledensity of approximately 57.70 main telephone lines per 100 people (as of 1999) [21].

In 1985, the Ministry of Communications was created. In December 1999, the Ministry merged with the Telecommunications Authority of Singapore to form the International Communication Development Authority of Singapore [27]. Privatization of the incumbent service provider occurred in 1993, with Singapore Telecom (Singtel) being created (Singtel is owned by Temasek Holdings, Central Provident Fund Board, and a distributed share ownership) and granted a monopoly on the provision of almost all telecommunications services [28].

In 1995, very limited competition was allowed in the resale of public switched telecommunications services, which allowed private operators to resell international direct dial, facsimile, and e-mail services. In 1997, TAS authorized the resale of leased lines. However, more advanced services, such as Frame Relay, remained within the province of Singtel's monopoly.

The Singapore market remained devoid of most competition, except in the most limited forms, until 2000. Accordingly, Singtel was responsible for providing domestic and international basic telecommunications services. However, unlike many other PTTs, Singtel did not suffer from lethargy. It began to completely digitize its network and increased its teledensity from 47.8 lines per 100 subscribers in 1994 to 54.3 lines per 100 subscribers in 1996 [21]. Of course, a certain amount of this upgrading could be attributed to the awareness by Singtel that the Singapore telecommunications market would open to competition in the near future, but some of that was also a direct response to customer demand.

The limited opening of the basic telecommunications services market was announced in 1996, with licensing to occur in 1998. Because this date was 7 years earlier than it had promised Singtel, the Singapore government had to pay Singtel substantial funds in compensation for lost income, pursuant to Singtel's concession contract [29]. Singapore committed to this limited market opening in its WTO commitment, with full competition to begin in 2007.

At first, much excitement surrounded the liberalization process, since many potential new entrants saw it as a means to service the many multinational regional headquarters located in Singapore. However, many new entrants for basic service were disappointed to learn that there would be strict foreign ownership limits (51% Singaporean owned). Initially, the Telecommunications Authority of Singapore announced that in 1998 it would award two licenses for basic services. However, the Telecommunications Authority of Singapore ultimately determined in 1998 that the market could not support such additional competition and ended up awarding a license to one other service provider, StarHub (a joint venture of ST Telecommunications, Singapore Power, British Telecommunications, and NTT). Two years from the award of the license (2000), StarHub was to be allowed to begin service provision (the 2 years between license award and service provision were to be used by StarHub to set up its operations). StarHub then paid $10 million, which was a fee comparable to that Singapore Telecom paid for its license in 1992 [30].

On the eve of service provision by StarHub, partially in response to the downturn in the Asian economies, the government made a surprise announcement and opened the market fully to competition, also eliminating any previous foreign ownership restrictions. Within 2 months of the liberalization announcement, the Singapore government had authorized 66 facilities-based service providers. This led to a lawsuit by Starhub against the Singapore government.

Competition brought entry by many global service providers. Specific impacts have included the increasing importance of Singapore as an Asian telecommunications and business hub. Competition, however, is still new in this market. It is clear that an effective regulatory regime must be put in place in order for Singapore to reach its full potential.

Vietnam: an underdeveloped market

Vietnam, with 76 million people and just 1.7 telephones per 100 inhabitants, has one of the lowest teledensities in the world [21]. Vietnam is substantially less developed compared with other Asia Pacific markets, such as

Singapore and Australia. A significant portion of the Vietnamese GDP is still raised from agriculture.

The government, however, began in the mid-1990s an effort to grow the telecommunications market, to such an extent that ITU has noted that it is the second fastest growing in the world. This statistic, however, is misleading, since almost all the growth is government initiated and it neglects the almost nonexistent basis from which telecommunications began to grow. Further, since Vietnam is not a member of the WTO, competitive pressures inherent in a lot of other markets are lacking.

The telecommunications market remains a monopoly, with only one service provider, Vietnam Posts and Telecommunications (VNPT), taking some very initial strides toward liberalizing its market. First, the government separated the regulator from the service provider in July 1995, when it created the Directorate General of Posts and Telecommunications. Second, the government has allowed limited participation of foreign companies in the mobile services market. However, VNPT is a key competitor in every sector of the telecommunications market; with the exception of these limited entries, no other direct competition exists in the market and little investment has been allowed in.

The government, however, has allowed companies to supply and install equipment in return for a share of the revenues for a period of between 5 and 15 years. Some of these projects have been successful in increasing the infrastructure level of Vietnam. For example, the Australian incumbent service provider, Telstra, has made substantial investments in VNPT's international and long-distance networks over the past decade [31]. In 1998, Telstra signed an agreement with VTI to construct and install an international earth station in Ho Chi Minh City in return for a percentage of international traffic revenues [32]. This project was successful enough that Telstra has continued its investment in other network projects, including the establishment of call centers. Several other similar projects have occurred both on the international and domestic networks of VNPT.

Another issue that VNPT is addressing is the differing standards within its own network. There have been estimated to be 12 different types of digital switches in the Vietnamese telecommunications network, which has created problems with interconnection. VNPT is trying to fix this problem by limiting the number of infrastructure vendors.

In addition, VNPT has set as a key priority the development of rural telecommunications networks. Many villages have never had any connection to the national telecommunications network. Extension to these villages has begun to take place [33].

The only competitors to VNPT to date are both state owned. These are the Military Electronic Telecommunications Co. and the Saigon Post and Telecommunications Joint-Stock Corporation. Both have announced significant plans for building out telecommunications services in Vietnam.

There is some competition in the mobile communications market. However, foreign investment is limited to a business cooperation contract, such that the foreign operator provides equipment and financing while VNPT manages the operation of the network. Accordingly, competition is not true, by any means. The data market is similarly curtailed and a VNPT subsidiary is the only player.

Accordingly, the Vietnamese telecommunications market is still essentially closed to entry. Despite pressures from other governments and industry, no plans have been made to open the market. However, it is interesting to note the continued effort of the government and VNPT to improve its network through creative forms of foreign investment. Unfortunately, the Vietnamese market still remains terribly underdeveloped and no true signs of improvement seem possible in the near future until the government opens its doors.

3.2.4 The Middle East

The Middle East is an extremely lucrative telecommunications market that many global service providers are anxious to enter to provide service, and much pent-up customer demand exists for new, innovative, and cheaper services. However, it remains stuck in only the very early phases of market opening, with little, if any, competition existing for any service in any market. In fact, recent ITU data show that the smallest amount of privatizations worldwide of incumbent service providers has occurred in the Middle East. Further, the Middle East is known for having the strictest limits on permissible foreign investment out of all regions, and only 10% of all Middle Eastern countries allows competition in basic service. Even traditionally liberalized services, such as mobile, remain relatively competition free, with only 30% of Middle Eastern nations having competitive mobile markets, as compared with 90% for European nations.

There are, of course, exceptions. Bahrain partially privatized its incumbent service provider, Batelco, in 1981 [34]. Jordan became the first Arab nation to sell a strategic share of its incumbent service provider in 2000, when it sold 40% to France Telecom [35].

Jordan

Jordan is one of the more liberalized Middle Eastern countries. It has a population of approximately 6.48 million people with a teledensity of approximately 8.34 per 100 main lines [21]. Through World Bank assistance, Jordan has been following a measured pace in restructuring its telecommunications sectors. Among the recent endeavors has been the creation of an independent regulatory body—first the corporatization and then the privatization of the incumbent service provider and then the adoption of new licensing rules for data communications.

Specifically, for basic services the Jordan Telecommunications Company has been the incumbent monopoly service provider. In 2000, 40% was sold to France Telecom and to the Arab Bank [35].

Prior to this occurring, the government created a regulatory agency, the Telecommunications Regulatory Commission, which has overall sector regulation. The Ministry of Posts and Communications has retained a policy role. The success of this process still remains to be seen.

The market for data services has been liberalized, but still facilities must be obtained from the incumbent. JTC has the basic services monopoly until December 31, 2004 [36]. After that date, there will be no restrictions on market access. It is still too early to comment on this process. However, what is clear is that many investors will watch this market with keen eyes to determine if true competition and a level playing field can be created in the Middle East's telecommunications markets.

3.2.5 Africa

Historically, Africa has continued to lag substantially behind the rest of the world, with the exception of the Middle East, in opening its telecommunications markets to competition. As of 2000, ITU data show that only 35% of its incumbent service providers had been privatized, as compared with 74% in the Americas and 63% in Europe. To date, movement is slow in all but a few markets.

South Africa

The South African market is of particular interest, since South Africa is often seen as one of the most important markets in Africa and often sets the trends for future markets. The Democratic Republic of Congo provides a stark contrast to the more liberalized South Africa.

South Africa has a population of approximately 42 million, with a per capita GDP of $3,215 [37]. Its teledensity is approximately 11.3 per 100 lines [21]. Accordingly, South Africa is substantially more economically developed than most of the other nations in Africa.

South Africa Telkom has traditionally been the incumbent monopoly service provider in South Africa. In 1997, a 30% stake in Telkom was sold to U.S.-based SBC Communications, Inc., and Telekom Malaysia Behrad [38]. As part of this process, Telkom was awarded the exclusive right to provide public switched telecommunications services for a period of no less than 5 years. In return, Telkom committed to modernizing its network and adding extensive phone lines.

All of these events were triggered by the end to apartheid, with 1994 bringing a new government, which recognized that universal access to telecommunications services was an important step in achieving social and economic development.

Among the first steps taken were the privatization of Telkom; the creation of a regulator, South African Telecommunications Regulatory Authority; and the adoption of a new telecommunications law. An important part of this process was also the establishment of the Universal Service Agency. Through the Universal Service Agency, telecenters were opened in rural areas, and the network expansion of Telkom was encouraged.

Subsequently, the telecommunications regime of South Africa has undergone some change. In the late 1990s, SATRA began to allow some competition in the form of Internet service providers and other VAS providers.

In addition, in 2000, the government of South Africa merged SATRA and the Independent Broadcasting Authority to create a new regulatory body called the Independent Communications Authority of South Africa (ICASA) [39]. ICASA has been given all the authority of the previous bodies. Policymaking is still reserved to the Ministry for Posts, Telecommunications, and Broadcasting.

In terms of WTO commitments, South Africa has made a commitment to end Telkom's exclusivity by the end of 2003 on facilities-based services. However, within this context, South Africa has the right to limit foreign investment to 30%. Currently, extensive discussions are being held on this issue.

Despite these positive steps, South Africa is still a fairly closed market, because Telkom has been able to use its dominant position to effectively block or hinder competition in key markets. As discussed earlier, Telkom has sued the regulator, arguing that it illegally authorized competition in the Internet service market. While this issue is pending in the courts, the overall tensions surrounding it have calmed. These tensions, however, have resurfaced in the area of the provision of Frame Relay services. The issue centers around whether or not Frame Relay service is considered a basic service that falls within Telkom's exclusivity. Telkom has argued that it is and has failed to provide any increased circuits requested by value-added network providers wishing to provide this service to customers. This has forestalled the creation of competition in this very important market sector. To date, this issue has been raised to the regulator and is still pending formal resolution.

As South Africa makes plans for its further market opening, many eyes will be carefully watching these developments.

Democratic Republic of Congo

The Democratic Republic of Congo is a market still in the very infancy of the telecommunications development. In 1997, there were only 21,000 main-line telephones in use. The GDP in 1999 was $31 billion [40].

L'Office Congolais des Postes et Telecommunications (OCPT) is government owned and provides all telecommunications services. The government has not issued any plans to privatize OCPT. There is no competition in wireline telecommunications, but there has been some initial market opening in the wireless services market.

Similarly, there has been no progress made on the creation of an independent regulator. Currently, the Office National des Posts et des Telecommunications du Zaire is the government entity with oversight of the telecommunications sector.

Unfortunately, this closed market structure is typical of many African countries, which have been hesitant to open their markets to private investment in the telecommunications market.

References

[1] Muller, Jürgen, "International Experiences in Competition and Regulation in Telecommunications Industries," 1994, http://www.oecd.org.

[2] Budde, Paul, "Decline and Fall: New Zealand Telco Regulation, Government Activity," *Communications International*, January 1, 2000.

[3] "One Regulator in Broadcasting: Summary of Changes Proposed," *The Guardian* (London), December 13, 2000, p. 6.

[4] As of 1999, the percentage of countries in the Americas that had completed some form of privatization was estimated to be 70%. ITU, "Americas Telecommunications Indicators 2000: A Decade of Reforms," 2001, http://www.itu.int.

[5] 47 U.S.C. § 1.

[6] 47 U.S.C. § 201, stating "All charges, practices, classifications, and regulations for and in connection with such communication service shall be just and reasonable."

[7] 47 U.S.C. § 254.

[8] *United States v. AT&T*, 552 F.Supp. 131 (D.D.C. 1982), *aff'd mem. sub. nom. Maryland v. U.S.*, 460 U.S. 1001 (1983).

[9] Latin America Regional Reports, Andean Group, "Venezuelan Telecoms Open Up in Earnest: CANTV's Monopoly Expired in November," December 12, 2000, p. 4.

[10] "Investment Opportunities: Honduras," *Institutional Investor, International Edition*, No. 9, Vol. 25, September 1, 2000, p. 6.

[11] "Central America Goes Ahead with Privatization: Costa Rica Moves More Slowly, Seeking Consensus," *Latin America Newsletters, Ltd.*, October 13, 1998.

[12] Wilson, James, "France Telecom Wins CTE-ANTEL," *Financial Times* (London), July 27, 1998, p. 17.

[13] http://www.siget.gob.sv/ (in Spanish).

[14] http://www.oecd.org.

[15] Office of Information Technologies Web site, http://www.ita.gov, "Information Technology Market Opportunities U.S. Small and Medium Sized Business: Highlighting Germany and the United Kingdom," accessed 2001.

[16] Jones, Phil, "Bundespost to Telekom: West Germany," *Communications International*, No. 6, Vol. 16, June 1989, p. 5.

[17] Telecommunications Act of July 25, 1996 (TKG) (*Federal Law Gazette*, Vol. 1, p. 1120), amended by Article 2, para. 34 of the Side Law to the Telecommunications Act (Begleitgesetz zum Telekommunickationsgesetz-BegleitG) of December 17, 1997 (*Federal Law Gazette*, Vol. 1, p. 3108) and amended by Article 2, paragraph 6 of the Sixth Law against Restraints of Competition of August 26, 1998 (*Federal Law Gazette*, Vol. 1, p. 2544).

[18] "U.S. Warns Deutsche Telekom Not to Block Market Access," *Capital Markets Report*, August 11, 1999.

[19] "You're Not Dialing Around Us," *Communications Today*, Vol. 5, Issue 222, November 15, 1999, available at 1999, WL 2852625.

[20] http://www.state.gov/r/pa/bgn/index.cfm?doicd=2852.

[21] http://www.itu.int/itudoc/itu-t/com3/focus/72404v7.xls.

[22] "MagyarCom Takes MATAV," *Market Trends, Eastern European & Former Soviet Telecom Reports*, January 1, 1994, available at 1994 WL 232708.

[23] "Hungary/Privatization-2: State Property 67%–70% Privatized," *Dow Jones International News Services*, December 21, 1995.

[24] Trade Practices Act of 1974.

[25] "Brave New World," *The Australian*, December 22, 1998, p. 38, available at 1998 WL 1823121.

[26] "Australia: LNP Exemptions, 1998," *Exchange Telecommunications Newsletter*, October 9, 1998, available at 1998 WL 9833303.

[27] Choy, Jorina, "Singapore's New Board to Drive Infocomm in Asia," *Asia Intelligence Wire*, December 13, 1999, available at 1999 WL 18295023.

[28] Hardy, Quentin, "Finger Pointing Follows the JR East Issue," the *Wall Street Journal Europe*, November 1, 1993, http://www.itu.int/itudoc/itu-t/com3/focus/72404v7xls.

[29] "RPTG/Singapore Telecom/Shrs-3: Inclusion in MSCIA Factor," *Dow Jones Asian Equities Report*, May 27, 1996.

[30] "NTT: NTT Consortium Is Awarded Service Licenses in Singapore," *M2 Presswire*, April 27, 1998, available at 1998 WL 1130888.

[31] "Telstra to Invest Additional 40 Million USD in Vietnam Telecoms," *AFX (AP)*, May 29, 1998, available at 1998 WL 13001079.

[32] "Australia: Telstra Has an Eye on the Future," *Asian Review of Business and Technology*, July 1, 1998, available at 1998 WL 1038578.

[33] Hanh, Ngo Hong, "Massive Telecoms Network Inked for Poor Central Regions to Spur Progress," *Vietnam Investment Review*, October 2, 2000, available at 2000 WL 9576627.

[34] http://www.batelco.

[35] "French Envoy Reviews 'Progress' in Jordan–EU Ties in 2000," *World News Connection*, December 20, 2000, available at 2000 WL 30997237.

[36] http://www.tre.gov.jo.abouttre.htm.

[37] United States Department of State, Bureau of African Affairs, January 20, 2001.

[38] http://www.telkom.co.za/company/profile/index.html.

[39] http://satra.gov.za.

[40] http://www.cia.gov/cia/publications/factbook.

4

Difficulties of opening markets to effective competition: domestic remedies

4.1 Overview

It is likely that over the next decade a significant portion of global telecommunications markets will be open to effective competition. As this occurs, each market will likely benefit from the impact that true competition brings to a marketplace, such as lower prices, greater choices, and more innovative services. In addition, creating an effective competitive marketplace will provide new entrants with the ability to compete on a level playing field. Unfortunately, until the concept of effective competition is embraced and implemented on a global basis, entrants will continue to face either partially or fully closed markets or markets where anticompetitive conduct has not been forestalled. In addition, in such markets, governments and consumers will not be able to realize the many benefits associated with competitive markets.

There are many reasons for the failure of markets to open to effective competition. In a large number of cases, this can be traced to a less than adequate regulatory regime. In other markets, it may be caused by anticompetitive conduct on the part of the incumbent(s) or other service providers with significant market power. In still other markets, causes may include economic or political conditions, which have encouraged the government to keep the doors to its markets either fully or partially closed. Each of these and other similar situations need to be remedied in order for a vibrant domestic telecommunications market to be achieved.

In order for governments to structure an effective competitive market and for service providers to be aware of pitfalls that may exist in markets that have not yet adopted and implemented this regulatory philosophy, it is important to identify the difficulties associated with market opening. Accordingly, this chapter focuses on the major reasons that governments fail to open their telecommunications markets to effective competition, including stumbling blocks to a successful liberalization process and failures of past privatizations. Further, this chapter suggests several domestic approaches for potential and existing service providers to try to obtain solutions to these problems. This chapter is complementary to Chapter 5, which addresses the international tools that are available to new entrants, consumers, other governments, and other interested parties to assist the opening of telecommunications markets to both de jure and de facto competition.

4.2 Why governments are hesitant to open their markets to effective competition

Despite the plethora of sound rationales and incentives to open a telecommunications market to effective competition that have been discussed throughout this book, many governments remain hesitant to commit to this step either in full or in part. Many of these governments base this hesitancy on their belief that a fully or partially closed market results in one or more of the following:

- Sufficient control over the telecommunications market to allow the government to fulfill goals, such as the ability to monitor communications, including content, among its nationals, and the ability to limit or expand access to telecommunications services in an effort to preserve government oversight;
- A unified or single government-owned telecommunications system that has been adequate and for which a change to competition would force replacement with a system whose operation and success is unknown;
- Predictable service levels;

- An opportunity to stall the introduction of competition until the government can recognize increased financial returns through a delay in the privatization and liberalization processes (i.e., fees for regulatory authorizations) until more stable or attractive economic conditions exist;
- Stable annual revenues associated with government ownership of the telecommunications network(s) that are often used to subsidize other goals, such as highway build-outs, health services, and the like;
- Steady employment by the incumbent service provider for a significant number of people.

Unfortunately, while these and others are valid beliefs, failure to open a market solely for such reasons is shortsighted. Governments with this mindset neglect to balance these and other rationales of limiting competition with the tangible benefits and incentives that are brought by the introduction and implementation of effective competition into the telecommunications market.

In fact, maintaining barriers to entry almost always creates more problems than establishing an effective competitive telecommunications regime. In markets where the government has retained barriers to market opening and failed, the following problems typically exist:

- The inability of service providers to obtain sufficient funding to build new and expand existing infrastructure to serve more of a country's population with services that are in demand by business and other users;
- Increasingly high non-cost-based prices for services, hence discouraging use by both business and residential customers or leading to the creation of haves and have nots in the market;
- Poor customer service, because the absence of competition often leads to lethargy in customer service;
- Hindering the creation of new jobs, including higher-paying, high-technology job, through increased numbers of service providers that are brought in by competition and the businesses that are attracted to the country through improved telecommunications services;

- A lack of new and innovative services, including the provision of high-speed data services, tele-medicine, tele-education, and other similar services;
- Failing to obtain sufficient funding and an influx of new technology to modernize the telecommunications infrastructure.

Nonetheless, many governments remain resistant to changing their regulatory structures, as well as the legal and political regimes, to allow the introduction and implementation of effective competition in their telecommunications market. The most explicit examples of such markets are those that have failed to open their markets to most, if not all, forms of competition or privatization of their telecommunications sector.

Key examples of such markets include Vietnam, Iran, Cuba, and Thailand. In each of these markets, all or almost all telecommunications services are provided by government-owned and government-operated telecommunications services and no or few competitors exist. These governments have often cited many of the reasons discussed earlier in this chapter for failure to open their markets. Similarly, in these markets, many of the problems readily associated with monopoly markets exist, including high prices and limited access to telecommunications services. While it is anticipated that growing pressure to create competitive services will increase in these markets, it is unlikely that this change will occur quickly or without many of the problems that typically accompany this changeover in more reluctant markets.

When compared with other recent markets that have begun to create effective competitive telecommunications markets, these closed markets clearly are not receiving many of the benefits that are associated with an open telecommunications market. For example, as discussed earlier, Brazil only recently opened its telecommunications market to competition. Yet, in the short time since it has introduced competition in the mobile telephony market (which is one of the earlier sectors the Brazilian government opened to competition), it has seen, among other results, the following:

- A fourfold increase in the mobile phone business between 1997, when competition was introduced in that sector, and 2000;
- A decrease in the same period of about \$106 for the activation of a mobile phone line (from \$136 to \$30) [1].

Many other examples exist and are discussed throughout this book. These include the benefits that are being recognized in Argentina, where November 2000 brought a complete opening to the market, and in the European Union, where each member state market has been opened fully for telecommunications.

4.3 Privatization and liberalization: what went wrong?

The discussion below centers on key issues associated with the failure to privatize or liberalize either in full or in part. In regard to a complete absence of privatization or liberalization, the causes tend to be fairly straightforward and generally include the lack of political will to change the status quo. In this situation, convincing a government to change its philosophy is fairly difficult and may involve expending substantially more political capital and a stronger overall effort than seeking changes to a more open regulatory structure. Accordingly, success in this area rarely occurs overnight and often requires potential entrants to expend considerable resources to encourage this result.

Many governments believe that the simple act of privatizing its incumbent service provider(s) or of issuing competitive regulatory authorizations will set the groundwork for a competitive telecommunications market. Unfortunately, while these are good first steps, they do not guarantee the creation of a truly competitive telecommunications market. These methods can fail either through inadequate regulation, oversight, or enforcement. Below is an examination of the impact of the future to privatize partially or to privatize fully, discussing such failures and some potential remedies that can be sought domestically.

4.3.1 Privatization failures

Major stumbling blocks to the creation of effective competition have been the failure of governments to engage in privatization of their incumbent service provider, engaging in a privatization that fails or produces bad results, or privatizing but failing to liberalize. In failing to privatize, the incumbent service provider is able to retain the advantages obtained by government ownership, while denying consumers the benefits that a privatized telecommunications service provider may bring to a market. In the second case of privatizing in an unsuccessful manner, this often results in inflated prices for services and poor service in the market. Finally, it is important to recognize that privatization alone will not bring competition. In fact, all that will be created is a privatized monopoly. Accordingly, liberalization and effective regulation are essential complements to any privatization process. Each of these scenarios is addressed in more detail below.

Failure to privatize

As discussed earlier, a substantial number of countries still have not engaged in any form of privatization, even in cases where some market liberalization has occurred. The major reasons for this failure include:

- A lack of political support to move forward with such economic reforms;
- A fear of unemployment that may accompany a privatization effort;
- The desire to receive large windfalls by forestalling the privatization process until the economics make more sense;
- A belief that a privatized service provider may not have the incentives to provide adequate telecommunications services to the country or may cherrypick to whom it provides services;
- The fear of managing or regulating a privately owned and operated service provider;
- The fear of losing out on the revenues that the government-owned service provider has traditionally brought into the government's treasury.

While these reasons are legitimate, each is manageable. For example, political support for a privatization effort can be garnered through the demonstration of the success of such efforts in similar economic and political regimes. Such was the case in Ireland, where the successes in other similar markets provided competition advocates with a firm basis to lobby for early market opening in Ireland. Further, in order to ensure that service is improved, the government, as a condition of licensing, can require the privatized entity to meet certain service commitments. This was done in recent market openings in Brazil, for example, where basic service providers were required, as a condition of their concession, to provide basic telecommunications services to rural and remote regions. In addition, fears associated with lost revenues for the government can be subsided by understanding that the increase in new competitors and businesses brought in by improved telecommunications services will increase the tax base and the overall financial health of the economy, often by attracting new businesses.

The results of failing to privatize have different impacts, depending on whether or not the market is being liberalized in a similar time frame. If the privatization process does not occur and the market is liberalized, the new entrants will have to compete against a government-owned and government-operated entity. This situation has often resulted in the following:

- Unfair political advantages being retained by the incumbent service provider;
- Either actual or the appearance of favoritism by the government toward the incumbent service provider;
- The lessening of competition, since potential new entrants are deterred by the concept of competing against a government-owned and -operated entity;
- Competition being forestalled to the extent the incumbent is able to retain any advantages and acts in an anticompetitive manner;
- Limited incentives for the incumbent to improve its services, infrastructure, and efficiency.

With regard to the impact of not privatizing and not opening the market to competition, the impacts are even more severe. First, essentially the failure to privatize means that there will be limited improvements in service, network expansion, and the like since no or limited new capital is available for these items. In many countries, governments are frequently hesitant to expand the funds necessary to build and operate a first-rate modern, reliable, and far-reaching telecommunications network. Accordingly, business as usual is expected to continue until new funding is provided to the incumbent or private investment is allowed in the service provider.

Of course, this failure to improve the telecommunications network may mean that the country will likely miss out on many new business opportunities, which rely on improved telecommunications services and infrastructure. Such business opportunities include call centers, banking, and the manufacturing sectors. For example, in regard to the banking industry, if high-speed data cannot be transmitted across borders using the telecommunications networks, it may be impossible to offer the worldwide standard of financial services typically associated with the bank, and the banking sector will locate its larger operations elsewhere.

Further, consumers will also likely continue to pay rates on a noncost basis for telecommunications services. In most cases, government-owned and -operated service providers do not have to abide by the same pricing principles that commercial operators do. Accordingly, in many cases the governments may structure their telecommunications fees to benefit a particular sector of the economy. For example, in most government-owned monopoly markets, the government-owned entity charges significantly less than cost for basic domestic telecommunications services. By doing so, the average consumer benefits and sees this cheap rate as a government benefit. In order to compensate, however, the government-owned operator will overcharge for international telecommunications traffic, which is typically provided to higher-end residential and business consumers. Such a price structure often works to deter new businesses to enter the market and creates an artificial pricing structure, which may not be financially viable over the long term.

Failed privatization

Although a government privatizes its incumbent service provider(s), the privatization may not always be successful. An unsuccessful privatization may result in poorer service than was previously provided, as well as a dramatic devaluation of the service provider's assets. Other negative results may include degradation of network assets, a decline in the introduction of new technologies and services, and increased rates for services. Such results may have severe repercussions in the market and directly negatively affect consumers and the economic conditions of the country.

Privatizations may be less than successful for several reasons. One of the most common reasons is that governments continue to retain some form of government ownership. This ownership can be as simple as a golden share to a controlling share of the service provider. In any event, continued government ownership provides at the least an appearance of government favoritism to the incumbent service provider. Often, this is more than an appearance and, in fact, often results in actual favoritism by the government to the incumbent service provider. A good example of such continued ownership is the fact that the German government still owns a share of Deutsche Telekom, the former government monopoly service provider. Because of this continued ownership, many new entrants and some other governments have claimed that Deutsche Telekom retains an unfair advantage in the market and has been able to stall the introduction of meaningful competition through such areas as interconnection negotiations.

In other situations, continued government ownership means that the investor is unable to make changes that would make the incumbent service provider a more viable commercial venture. Such changes could include rebalancing rates to more accurately reflect costs, eliminating excess personnel, refocusing energies on other services, and making changes in management, among others. Unfortunately, when the government retains a say in the operational aspects of the incumbent, many of these fundamental changes may not be able to be made.

Another reason for failed privatizations includes the lack of government oversight and effective regulation over the privatized entity. Failure of the government to adequately police the privatized entity may lead to numerous problems, including the privatized entity failing to meet the obligations agreed upon in the privatization process. Such obligations typically include increasing service to a larger percentage of the population, furthering its network build, modernizing the network, and lowering rates, among other things. An example of this situation has been the privatization of Romtelecomm, where 35% was purchased by OTE [2], the Greek incumbent service provider. Romtelecomm is also owned in part by the GTE Corporation. Unfortunately, many of the benefits that were anticipated with the privatization of Romtelecomm have not materialized. In particular, the partially privatized Romtelecomm has failed to meet expectations of build-out of its network to underserved areas of Romania. This has resulted in dissatisfaction among consumers and was raised as an issue in the highly contested presidential political campaign in 2000.

Another major reason for a failed privatization effort is the government overvaulting and bidders overpaying for the service provider. In this situation, the privatized operator may not have sufficient capital in order to meet the service commitments that were expected of it during the privatization process or fulfill the expectations of the government and consumers, as well as itself. If the 2000s see further economic turndown, the world may see increasingly failed privatizations for this reason.

Another path to failure involves the government privatizing its incumbent service provider(s) but failing to liberalize the market. This approach allows the incumbent to retain unfettered dominance of the market. Accordingly, the incumbent retains its monopoly attributes, including benefits associated with this status and relevant attributes. It should be noted, however, that privatization without liberalization is still a better approach than no privatization at all. Such an effort still provides the government with the ability to extract concessions by the service provider to improve service and network. Nevertheless, governments must be careful not to rest on their laurels. Many problems exist with this approach. An example of this is South Africa, where privatization and only very limited liberalization has occurred. Because the basic service market has been retained as a monopoly, Telkom South Africa, the incumbent, has been able to discriminate against new entrants for value-added services, such as Internet, and continue to charge customers exorbitant rates for services [3].

4.3.2 Liberalization

Another major stumbling block to governments instituting competitive telecommunications markets effectively is that many governments either purposely or by happenstance stall liberalization or only partially allow liberalization to take hold. As many examples demonstrate, not every liberalization process runs smoothly or results in the creation of a competitive market for telecommunications services.

There are many reasons for liberalization attempts to fail or not reach their actual goals. First, quite a few efforts at liberalization have led to little or no interest on the part of investors. This results in a government failing to be able to issue, at least initially, the expected number of authorizations it had estimated. Quite often this kind of failure means that there is something wrong with the liberalization process. This can include:

- A government overvaluing the resources it is licensing or auctioning;
- The failure of a government to create a regulatory regime that provides adequate comfort to the potential new entrants;
- Changing economic conditions either in the country or globally that impact the financial attractiveness of the market to investors.

Insufficient interest in the market may permit incumbents or other providers to retain or gain a larger market share than originally contemplated, hence having control of a specific market sector. This often leads to higher prices for consumers and fewer choices.

Another problem often faced with the liberalization process is that the ultimate tangible results of the process are different from those that are expected or announced or take a longer time than anticipated. Many times governments will expect that opening their markets fully or opening limited sectors to competition will result in vast consumer benefits, such as substantially lower rates or improved infrastructure and network reach. Unfortunately, that is not always the case. Often, for the benefits of liberalization to be recognized, some time must pass between evaluation and the opening process. In other cases, service providers must have incentives to meet these goals. For example, in order to ensure greater coverage for telecommunications services, many governments have instituted a universal service policy, which requires individual basic service providers to contribute funding.

Second, in some cases, there may be other issues impacting the success of the liberalization process. For example, the regulatory regime, although fair and transparent in the authorization phase, many not be fully fleshed out for actual operations, making actual operations difficult. Further, interconnection regulations may not be fully developed or enforced so that valuable resources of new entrants have to be spent on this issue, as opposed to network build-out.

Other reasons may include the effects of political or economic instability in that market. Accordingly, new entrants may be hesitant to invest additional monies in the market or subject personnel to volatile political situations. In these cases it is imperative that the government try to work with service providers to institute safeguards, such as rules on the expatriation of profits, to provide a minimal comfort level to the service provider.

Another major issue is the failure by the government to adequately define what areas constitute competitive services. This especially creates problems in markets where the incumbent retains a basic service monopoly with only a few services being made competitive. For example, in Colombia, where the government retained for several years a monopoly for basic services, but allowed competition in value-added services, there was disagreement over what constituted a value-added service. Telecom Colombia, the monopoly basic service provider, believed that certain high-speed data products fell exclusively in its domain, while many of the value-added service providers believed otherwise. The government, however, never stepped in to clarify this situation, which ultimately resulted in a retarded market for these services.

A further issue is the failure of a government to curb the anticompetitive abuses of incumbent service providers. Often, these abuses prevent a competitive market from being fully developed. For example, in Mexico, Telmex has been able to act in an anticompetitive manner to block competition from truly developing in the basic service market [4]. Unfortunately, despite major efforts by new entrants, the Mexican government has consistently failed to curb the actions of Telmex. Examples of anticompetitive actions by Telmex, which were seemingly endorsed by the Mexican government, include:

- Anticompetitive restraints on cross-border communications services. Despite a specific request by new entrants in the Mexican market, the regulator refused to authorize the provision of cross-border telecommunications services independent of the current accounting and proportionate return system. This has perpetuated Telmex's control of the international market.
- Increasing interconnection rates. Instead of lowering basic interconnection rates, a ruling by the regulator actually raised the rates that Telmex may charge, in contrast to worldwide trends. These rates are not cost oriented.
- The regulator has never enforced Telmex's legally mandated cost imputation and accounting separation obligations. Accordingly, Telmex has been able to put a price squeeze on long-distance competitors [5].

4.4 Domestic remedies

This section focuses on the myriad domestic remedies available to service providers, consumers, and other interested parties when the privatization or liberalization processes fail or are inadequate. These remedies are general in nature, with each adaptable to the specific market, situation, and user.

It is important to remember that each form of domestic intervention is resource intensive. Unfortunately, it is nearly impossible for the casual participant to be successful at achieving results. Accordingly, many consumers and service providers will pool resources and participate under the guise of trade associations or in informal groups. In this manner, there is increased strength, because of the increased number of participants and a reduction of individual resources that need to be expended. Of course, the richer advocates are able to participate both on an individual basis and with others.

In terms of formal trade groups, chambers of commerce are often one of the most active in many countries, because of the large number of major companies that comprise its membership. Other key trade groups include smaller associations that are made up of a set group with similar interests, such as the Value-Added Network Service Provider Association in South Africa, which is comprised of value-added network service providers.

In regard to less formal coordination, many times companies or users with similar interests will lobby in informal groups or at least coordinate their actions to obtain similar results. In this regard, key government decision makers are hearing consistent views from a larger number of parties than would be possible if such coordination did not take place.

No matter what remedy is pursued, it is important to remember that any remedy that is sought will generally take time to achieve. Unfortunately, since telecommunications regulation is legally based, change takes time. Accordingly, long-term wherewithal is imperative for new entrants. This generally means that new entrants should prepare to have a financial reserve to take into account the resolution of legal, regulatory, and policy changes that will need to be made in order to achieve their planned operational results.

However, such remedies may be easier to achieve with a set strategy. Accordingly, early on the advocate should establish its strategy, recognizing that it may need to evolve as its efforts for change go forward. In all cases, it is important to expand its arguments for change outside of one that is solely focused on the welfare of the advocate. In general, it is key to determine the interests and mandate of the government official or body being addressed, the mandates required by law, and craft arguments accordingly. For example, many telecommunications laws have certain mandates that the execution and implementation of the law are supposed to meet. For example, most recent telecommunications laws have requirements that benefit consumers, universal service extended to all portions of the country. Accordingly, if an advocate service provider is going in to meet with the government to argue for a change of a rule or greater enforcement of the market, it would probably be helpful to craft a strategy that in part demonstrates to the government why that will serve the interests of the law. In addition, it is important to remember that most governments are looking to bring additional capital into a marketplace. Accordingly, bringing in economic arguments is often of help. The local consultant the advocate utilizes can generally assist in identifying key points that are important to the audience being lobbied, as well as what to avoid.

In each case, it is extremely helpful to hire local attorneys and consultants, who can assist a company or consumer group to work within the confines of the particular country's legal, regulatory, and political framework. Many governments are hesitant to work with foreigners who do not employ such local assistance.

With this background in mind, it is helpful to understand the general forums that are available to advocates seeking domestic remedies. The following discusses how an advocate can appear domestically for assistance in remedying market access issues.

4.4.1 Legal

Often, in regard to privatizations and initial liberalizations, the constitution of a country or basic telecommunications law must be amended in order to allow either to occur. This requires action on the part of the country's legislature and often its executive. These kinds of actions are often slow to occur and involve a large amount of political capital for the relevant actors to move forward. In this regard, lobbying needs to be quite extensive and it is generally necessary to also provide draft laws that the advocate would like to see enacted. This will often provide the relevant politicians with a baseline with which they can seek support and a point with which the advocate can seek to obtain support.

A good example of where this change was required was in Brazil. First, in order to open the market to any but the most limited form of competition the Brazilian government had to amend the constitution. Once this was accomplished, the legislature had to pass several new laws, which would govern the telecommunications market. This process took several years to complete.

4.4.2 Regulatory

In general, one of the most important areas of recourse is generally regulatory. However, this option usually only works when there is an independent regulator established in the market. In markets where the government service provider is also the regulator, obtaining relief will be extremely difficult.

In those markets where an independent regulator exists, there are two types of regulatory actions available. The first type is informal and the second type is formal. With either type, it is extremely helpful to locate a sympathetic ear within the regulatory sector who will help champion the cause that is at issue.

In regard to informal action, this would include having informal discussions with the regulator, including the staff and senior officials, to explain the issue at hand and to offer proposed solutions. Informal action, however, will rarely lead to any concrete results except laying the groundwork for more formal action. However, if there is no immediate problem at hand, it may provide the regulator with some ideas on how to handle issues down the road and may allow the advocate to identify a regulatory champion. This regulatory champion can help when the advocate is seeking more formal action.

In general, in obtaining concrete action, more formal recourse is usually required. This type of action generally includes participating in rulemaking proceedings or filing formal complaints. However, in many cases, the regulator is not required to act by law, may act extremely slowly, or may not be inclined to rule in the favor of the advocate. This is where the regulatory champion can play an important role. In addition, as discussed below, intervention by other government agencies can also assist in furthering the cause of the advocate.

If an unsuccessful result is achieved, it is important to understand the recourse available to the advocate. This often includes having to re-petition the regulator for relief before seeking judicial intervention. Such appeal processes are often time-consuming, and in many cases the regulator may not be required to act on the appeal. In some cases, however, judicial intervention is directly available. In other cases, recourse is less clear. This is where an in-country expert is extremely valuable and can help direct the advocate to the most efficient form of appeal possible.

If a favorable result is achieved, enforcement of the decision is the key to success. This may be another important place where the regulatory champion may be able to assist in encouraging the regulator to take the actions required by the decision. In addition, the involvement of other government agencies or officials may be of help if the regulator is unwilling to act. This has been somewhat effective in Mexico, where the regulator, COFETEL, has been largely ineffective [6]. In this case, the Mexican antitrust agency has stepped up with some assistance to new entrants in enforcing the Mexican regulatory regime.

4.4.3 Other government agencies

Other government agencies are also becoming increasingly important in establishing the legal, regulatory, and policy regime governing telecommunications. As already discussed, antitrust authorities are becoming more visible in regulation of the telecommunications market. A good example of this is the European Union's Directorate General IV, which has been quite active in the merger review process. Other agencies, such as trade authorities and government bodies with jurisdiction over international issues, such as foreign trade ministries, are also becoming increasing visible. All of these agencies, including local ones, may be able to intercede on behalf of the advocate to obtain results from the regulator or with the lawmaking bodies, or may be able to act directly.

The growing importance of antitrust agencies is demonstrated by the rejection of the WorldCom, Inc./Sprint merger in 2000 by the European Union's Competition Directorate [7]. In that case, two American companies had to seek European Union approval for their proposed merger under Articles 85 and 86 of the Treaty of Rome and the European Union's Merger Guidelines. The European Union, in an unprecedented decision, determined that the merger would be anticompetitive, even after the parties withdrew their application from the EU's consideration.

4.4.4 Judicial

In some countries, advocates may be able to bypass the political bodies that generally govern the day-to-day aspects of telecommunications and proceed directly to the judiciary. Although this is a rare option, it is an important one to examine. Often, in extremely political situations, the judiciary may be the most independent body. Judicial remedies, while of a binding nature, generally take a long time to be released and enforced. Accordingly, for short-term relief, this may not be the best path to follow.

Perhaps the most impressive use of the courts to open telecommunications markets to competition was by MCI Communications Corp., discussed in Chapter 3. As discussed, MCI Communications Corp., when turned down by the FCC for a long-distance service authorization, sought redress from the courts. The courts, interpreting the governing Communications Act of 1934, in conjunction with the Department of Justice, determined that the FCC must allow competition.

4.4.5 Political bodies

In many cases, political remedies may be the most effective. This can include lobbying members of the government's executive and legislative bodies. Depending on the relief being sought, informal or formal action can be taken. In terms of informal action, this may involve a simple letter of inquiry from a legislator to the regulator on the matter at issue, which demonstrates a high-level government representative's interest. In terms of more formal actions, hearings may be opened or laws proposed and passed.

An excellent example of a political success in obtaining a market was by the founder of PanAmSat, a satellite company. Renee Anselmo, a millionaire from the television industry, believed that a good market existed for international satellite services. However, the U.S. government (including the FCC) argued against allowing anyone but INTELSAT, an international treaty organization, to launch and operate international satellite systems. In response, Mr. Anselmo launched a massive and aggressive lobbying campaign in all branches of the U.S. government. Mr. Anselmo ultimately won support for his position to open the international satellite services market and today PanAmSat, which is now owned by Hughes, has the largest fleet of international communications satellites in operation.

4.5 Next steps

Many times seeking remedies in domestic forums is sufficient to obtain adequate relief from problems a new entrant is facing in the market or provide impetus to a government to either privatize or liberalize its telecommunications market. However, there are other times where such actions are not sufficient and advocates must look for additional support. An important area to obtain support is through the use of intergovernment communications on a bilateral and multilateral basis. Chapter 5 discusses this avenue in depth.

References

[1] http://www.anatel.gov.br.

[2] "Greek Telecommunications Move from Monopoly to Competition," *International Herald Tribune*, January 11, 2000.

[3] Horowitz, Robert B., "Telecommunications Policy in the New South Africa: Participants, Politics, and Sectoral Reform," *South African Journal for Communication Theory and Research*, Vol. 23, No. 2, 1997.

[4] Turner, Oliver, *Mexican Telecommunications*, accessed January 2001, http://www.takingstockllc.com.

[5] USTR Press Release, "USTR to Request Consultations with Mexico Regarding Telecom Trade Barriers," July 29, 2000, available at http://www.ustr.gov/releases/2000/07/00-57.html.

[6] Congressman M. Oxley, Letter to USTR C. Barschefsky on Opening the Mexican Telecommunications Market, March 10, 2000.

[7] McDonald, T., "EU to Block WorldCom/Sprint Merger," *E-Commerce Times*, June 21, 2000.

5

International market opening and leveling tools

5.1 Overview

This chapter complements Chapter 4 and focuses on international market opening and leveling tools that are available to entrants in global telecommunications markets. In today's global economy, service providers and their investors often do not originate from the country they are planning to enter or where they are operating. In this situation, a government, through its domestic process, may be less receptive to these groups than similarly situated nationals of the country. Accordingly, it may be helpful to seek the assistance of the government of origin of the service provider or investor or another country where the service provider or consumer is already established to receive assistance in encouraging the creation of a competitive telecommunications market or to seek a specified relief from a problem that has arisen in the country's telecommunications market. This assistance can take place through myriad forums, which are discussed below. In addition, at times the entrant itself may be able to play a direct role in seeking and obtaining such relief either through participation in a country delegation to an international meeting or, where permissible, through its own membership or participation. Another option is to work through the existing international trade laws.

Because of the intensive resources involved in seeking such assistance internationally, the entrants that are best able to pursue this course of action tend to be the larger, multinational corporations, such as AT&T, WorldCom, Inc., Verizon, France Telecom, British Telecom, and Sprint. Typical resources required to ensure successful resolution of entrant and competitors' issues include, but are not limited to, funds for extensive lobbying efforts, the payment of fees to international organizations for membership and to attend conferences, and funding for international travel. Large service providers are often assisted in their advocacy by their larger consumers, such as Proctor & Gamble or Citibank. These consumers have an important stake in the competitive landscape because their day-to-day operations are global in nature and involve vast use of, and expenditures on, telecommunications services. Smaller entrants and consumers are less likely to take such actions but may be the beneficiaries of the actions of others. In addition, these parties may be able to play a role in obtaining relief on narrow issues of concern. Such actions may include joint lobbying efforts, participation in trade associations, and raising the issues when the opportunity presents itself.

A key consideration for an entrant before pursuing an international solution is the length of time involved in reaching a resolution. Because of the nature of the international process and international relations, results are rarely instantaneous and, in fact, may take years to achieve. The length of time to achieve results is dependent on several key factors, including the willingness of the parties to achieve resolution of an issue, the process in which these results are sought (e.g., multilateral versus bilateral negotiations, use of a formalized dispute resolution mechanism, unilateral actions), and the political importance the governments involved place on achieving a successful resolution. In addition, quite often the results achieved are the result of direct compromise, so they may not be the entrant's ideal solution. Further, although success may appear likely throughout the process, it may be that the issue being negotiated is traded by the respective governments for the resolution of another issue, perhaps totally unrelated, that the government deems more important to its interests to resolve. In such cases, the private sector advocate may be left without recourse in the international, or any other, forum.

A benefit of achieving resolution of a contentious issue through a negotiated international process is that the agreed-upon resolution is often more likely to be implemented because of the international political pressure that is placed on the subject country to comply with its international obligations. Failure to live up to one's international obligations can have a significant impact on a country in many areas related to telecommunications, as well as those unrelated to telecommunications, including financial assistance for areas such as health, military, and educational programs. Often a country that fails to abide to an agreed-upon commitment may find it harder to get international aid or other assistance in the future from the disputing country or its allies.

It is also important to keep in mind that another possible form of international resolution is through unilateral action by a government. Unilateral action generally is imposed when an action by the other government is deemed egregious and normal settlement mechanisms have failed to resolve the issue. This type of action is fairly common in heated trade disputes. However, in only selected cases does this action achieve the desired results, because such an approach is often seen as an affront to another country's national sovereignty.

5.2 The use of international organizations to obtain results

As discussed, a key method of obtaining liberalization results is through participation in international organizations that have as their full or partial mandate to address telecommunications issues. Although there are several organizations that have such a mandate, the key international organizations in this area are the International Telecommunication Union, OECD, and the WTO. Below is a discussion of each of these institutions and the role they can play in assisting in opening a market or leveling the playing field in a particular country or region. In addition, there is a brief discussion on the World Bank, which plays an important role in shaping the regulatory structures of countries because of the commitments associated with the receipt of World Bank funding.

5.2.1 The International Telecommunication Union

An international organization that has taken on growing importance in creating a level competitive telecommunication market is the International Telecommunication Union (ITU), a treaty organization and an arm of the United Nations. The ITU has three primary missions:

1. The establishment of international technical standards and frequency management;

2. The promotion of the development of the telecommunications sector in lesser-developed countries;
3. The formulation of broad international policy goals on telecommunications issues [1].

The ITU is comprised of approximately 189 member states and also has over 650 sector members, including equipment manufacturers, service providers, broadcasters, and other international and regional organizations [2]. While sector members can participate in most ITU activities, the treaty-making power of the organization is limited to member states.

The ITU is made up of three primary sectors in which to carry out its mandate. The important sectors, which have a direct impact on the creation of an effective competitive telecommunications market, are the Telecommunication Standardization and the Telecommunication Development Sectors.[1] The Telecommunication Standardization Sector coordinates the international telecommunications standards-setting activities, which are turned into ITU-T recommendations and which address key telecommunications policy issues [3]. The Telecommunication Development Sector has the mandate of promoting investment in and the implementation of telecommunications infrastructure in developing countries through a variety of means, employing techniques such as training programs and financing strategies [4]. In general, both forums are useful when attempting to influence broad policy issues on market access and entry.

The ITU is an excellent body in which to participate if the change being sought is long term and impacts many countries. A good example is the ongoing work of the ITU-T on accounting rate reform. Accounting rates are the monies paid by international service providers to carry each other's telecommunications traffic on a half-circuit basis. Traditionally, primarily because of the monopoly structure of the telecommunications industry and the level of technological development of the associated telecommunications networks, these costs have been very high. However, with the onset of competition, these costs have dropped dramatically. The ITU-T has been instrumental in providing increased pressure on completing this reform process [5].

1. The third sector, the Radiocommunication Sector, is concerned with establishing technical characteristics and operational procedures for wireless services. It also bears substantial responsibility for radio frequency coordination and management.

In addition, the ITU has created a World Telecommunications Policy Forum (WTPF) where member states and sector members can discuss and exchange views and information on emerging telecommunications policy and regulatory matters arising from the changing telecommunications environment [6]. Although the WTPF is not authorized to produce prescriptive regulatory outcomes or outputs with binding force, it is enabled to prepare reports and, where appropriate, opinions for consideration by members and relevant ITU meetings [6]. Previous WTPF topics have included internet protocol telephony, trade in telecommunications services, and global mobile personal communications services.

Another key area that the ITU handles is the creation of global technical standards for telecommunications networks and equipment. Many equipment manufacturers and service providers participate in these discussions, which often result in treaty agreements or voluntary standards. In many cases, the adoption of these standards directly impacts the ability of service providers to offer telecommunications services using the technology they believe is most suitable [7]. Figure 5.1 shows a Frame Relay switch, the type of equipment that might be directly impacted by the ITU standards-setting process.

However, like most international treaty organizations, the ITU is a poor forum for achieving a fast resolution of any issue. It is a deliberative body, where study is an important part of the course of most actions. Accordingly, many companies look at it as a long-term solution but not a forum in which to obtain any immediate relief [8].

5.2.2 The WTO

As previously mentioned, the WTO Basic Agreement on Telecommunications is a cornerstone for establishing certain key global telecommunications regulatory principles and for seeking redress over the long term of particularly heinous anticompetitive conduct in the markets of WTO signatories [9].

To date, the WTO has been utilized in primarily four ways. First, it is used as a point of pressure by countries seeking WTO membership in order to obtain liberalization and market access commitments. For example, when China sought entry into the WTO, the United States put pressure on it to improve access to its telecommunications market for foreign companies. Over the course of these negotiations, China improved its initial offer largely because of the pressure by service providers in other

Figure 5.1 Frame Relay switch.

countries to seek concessions from China before allowing it to reap the benefits of membership in the WTO.

Second, during the negotiations for both the 1998 WTO Agreement, and the upcoming round, governments continuously place pressure on other countries in the negotiation process to improve their offers to open their markets and create effective regulatory regimes. As part of this process, service providers often voice concerns about individual countries' commitments and broad principles, which they believe should be made part of the WTO commitments and principles to their member governments through both formal and informal mechanisms. A good example of this was the much-improved offer that was ultimately agreed to by the Canadian government during the 1998 round of the WTO Agreement negotiations. Members of the telecommunications industry in the United

States made a very compelling case to the U.S. government on the importance of obtaining an improved commitment from Canada prior to conclusion of the WTO Agreement in order for them to be able to provide services in Canada and between Canada and the United States, as well as elsewhere. In part because of this advocacy, the United States had added incentive to negotiate with Canada to improve its offer.

Third, once the commitments are agreed to, a government or private company may use the existence of the commitment as a pressure point to force a government to comply with its treaty obligations [10]. For example, failure of a government to meet a liberalization milestone or a principle can be addressed through informal negotiations, which point out that the country is failing to live up to its international commitments. Such discussions have some force, because of the threat of the dispute settlement mechanism that accompanies the WTO and because of the international pressure to conform to a country's commitments.

Finally, as in the case of Mexico, if noncompliance with a commitment occurs, a government has the opportunity to bring it up through the WTO dispute settlement process. To date, this process has not been fully utilized to address anticompetitive conduct in telecommunications markets, so its success rate is unmeasured [11].

In terms of the substantive part of the WTO Agreement, as detailed in Table 5.1, the agreement addresses individual countries' commitments on market access, including licensing requirements and foreign ownership restrictions, national treatment, and establishing nonbinding regulatory principles. The nonbinding regulatory principles are contained in a docu-

TABLE 5.1 WTO Overview

Market access and national treatment
Foreign investment
Regulatory principles, including:
◆ Competitive safeguards
◆ Interconnection
◆ Transparency
◆ Independent regulator
International dispute mechanism for failure to meet commitments

ment called the Regulatory Reference Paper, which can be modified by each country as it sees fit. The Reference Paper addresses several key principles, including:

- The establishment of an independent regulator;
- Nondiscriminatory, cost-based interconnection;
- The establishment of competitive safeguards;
- The creation of a transparent regulatory process.

Although nonbinding, the importance of the Regulatory Reference Paper should not be underestimated. It provided the first public commitment by many closed or newly liberalized markets to adopt a regulatory structure that fosters competition in the telecommunications market.

In terms of the individual commitments, the agreement addresses two basic principles. The first is transparency. Under the Basic Agreement on Telecommunications, each country must publish all key market opening information, including all notification, registration, or licensing requirements, and any other forms of approval, before a foreign service provider can do business lawfully in a member country. Another key principle is a limit to using licensing requirements as barriers to trade. Specifically, the WTO requires that licensing requirements do not constitute unnecessary barriers to trade. Under the national treatment provision, governments are required to grant market access to operators from a WTO member country (even if the country has made no commitments) on terms no less favorable than the terms applicable to operators from any other country.

The WTO Agreement has proven to be an effective tool over the long term to help curb anticompetitive abuses in the global telecommunications market and to encourage further liberalization. In terms of encouraging market access, by formalizing individual commitments in a binding multilateral agreement, there is express pressure on individual member states to comply with their commitments. Further, the regulatory principles encourage governments to meet minimum standards required for the creation of competitive telecommunications markets. These standards provide a firm basis for discussions between governments, as well as between governments and service providers, on issues that arise that are not in line with the individual commitments or the regulatory principles. Similarly, the national treatment provisions provide a way to ensure that governments do not act in an arbitrary manner such that the nationals of their own country are favored in market access over the nationals of another country.

As discussed, the WTO also has an enforcement mechanism through dispute settlement for failure of a member state to live up to its obligations under the WTO Agreement. To date no formal actions have started, but some preliminary uses of the dispute mechanism have been explored. In July 2000, the United States announced that it was going to request WTO consultations with Mexico for failure to implement its commitments under the WTO Agreement on Basic Telecommunications [11]. If this case proceeds to dispute settlement, it will be a test case of the use of this forum to reach a binding decision. The three issues under consideration are:

- Lack of effective control over the incumbent and former monopoly, Telmex, which has used its market position to curtail competition;
- Failure to ensure timely, cost-based interconnection, which would permit Telmex's competitors to connect to Telmex's customers to provide local, long-distance, and international services;
- Failure to permit alternatives to an outmoded system of charging U.S. carriers above-cost rates for completing international calls to Mexico.

To date, there has been some movement toward settlement of the U.S. dispute that may lead WTO action to be forestalled. However, it is likely that the threat of bringing the U.S.–Mexico dispute to the WTO for settlement may have helped to entice the Mexicans in seeking a solution.

5.2.3 OECD

OECD provides its 30 member states (whose economies produce approximately two-thirds of the world's products and services), all committed to a market economy and pluralistic democracy, with a forum to come together and develop policies designed to benefit the world economy through the sharing of their different ideas and experiences [12]. The mandate of OECD is primarily to perform economic research on various aspects of its members' economies and policy concerns. It has been addressing telecommunications issues over the past several years in an effort to encourage members to adopt and implement public policy in a uniform manner through international treaties.

To this end, OECD has held several international conferences focusing on international telecommunications issues. These conferences have addressed many policy matters, including access and use of infrastructure, minimizing regulatory uncertainty, and realizing the potential of electronic commerce. Although not a policy-setting body itself, OECD has great influence over its members in attempting to achieve common policy goals in order to create a level playing field in telecommunications. Accordingly, policy issues of broad concern, in areas such as interconnection, are often raised and addressed in this forum.

5.2.4 The World Bank

Another important forum for resolving key regulatory issues is the World Bank. The World Bank has played a prominent role in the reform of the global telecommunications market in developing economies. Initially, the World Bank was involved in financing the expansion of telecommunications infrastructure by the incumbent service providers. However, in the 1990s, the World Bank began to shift its direction. While continuing to fund such efforts, the World Bank has begun to seek commitments from the government it is funding in a host of areas, including privatization and liberalization efforts. In many of the markets where the World Bank has provided funding, such as Ghana, broad steps have been taken in opening the market to effective competition. Influence into this process is best achieved by meeting with the key personnel in the country or its telecommunications sector group [13].

5.3 Regional organizations

Another integral method of obtaining results in issues that impede competition is the use of regional organizations. Each region has many key organizations, including forums on telecommunications liberalization and policy. An example of several of these important regional organizations in regard to market access and creating a level playing field are:

- The Inter-American Telecommunications Commission (CITEL), which is a part of the Organization of American States, has as its primary mission the promotion and facilitation of the development of telecommunications in the Americas, in an effort to improve the overall development of the region. As part of this mission, CITEL focuses a significant portion of its time on telecommunications policy and development issues [14].

- The European Telecommunications Office, whose mandate is to support the establishment of new regulatory regimes for liberalized telecommunications markets and support the harmonization of telecommunications regulations [15].
- The African Development Bank provides financial and technical assistance for the establishment, expansion, improvement, and integration of public telecommunications systems in Africa. It has as its primary focus infrastructure development and increased access to telecommunications services [16].
- The Telecommunications Regulators Association of South Africa is an association of member states in the South Africa region. The goal of the organization is to increase communications and coordination between regulatory authorities in the South Africa region in order to improve the economic and social well-being of their nations through telecommunications sector reform [17].
- Asia-Pacific Economic Cooperation (APEC) was established in the late 1980s in response to the growing interdependence among Asia-Pacific economies. APEC is the primary regional vehicle for promoting open trade and practical economic cooperation in the Asia-Pacific region [18]. A key part of APEC is APEC TEL, which is the group that addresses telecommunications issues. Key recent developments include formulating guidelines on interconnection and working on bridging the digital divide.

Similar to the international organizations, these regional organizations can be of assistance in seeking redress to broad policy issues that are specific to a particular region of the world. Quite often, these smaller organizations or more focused organizations are able to address more specific problems than the larger worldwide forums. While governments continue to be the dominant force in these organizations, private sector participation is permitted in many of them, and often encouraged.

5.4 Bilateral and other multilateral trade agreements

Another effective method to obtain redress and results in market opening are bilateral and other multilateral trade agreements. In regard to bilateral agreements, these may be specific to telecommunications or may address broader areas of concern. A good example of successful bilateral agreements are the recent agreements that the United States has entered into with its neighbor Mexico over spectrum coordination issues. These agreements address specific market access concerns.

Multilateral agreements may also be specific to telecommunications issues or cut across broader issues. A good example of the latter is the North American Free Trade Agreement (NAFTA), which covers myriad issues, including telecommunications, transparency, monopolies, value-added services, and the like. Disputes under NAFTA are required to go to dispute settlement under the mechanisms provided for in the agreement [19].

Negotiating agreements such as NAFTA and the U.S.–Mexico bilateral agreements provides a good basis for countries to seek to address market access issues. Further, once the agreements are in place, they provide an additional pressure point for governments to utilize in seeking resolution of issues that conflict with the treaty. If, unfortunately, resolution of these issues cannot be reached through more informal means, the dispute settlement provisions of the agreements are available.

5.5 Intergovernment relations

Another important avenue available to service providers is to work within their government to garner support for approaching other governments in which the service provider operates to achieve desired changes. There are both formal and informal mechanisms to obtain this, and often this approach may be pursued on both paths.

In terms of more formal mechanisms, in public proceedings regarding the implementation of international obligations, such as trade agreements, governments may seek comments from interested parties about whether the obligations under the agreement are being met. Based on this input, the government may determine to formally discuss its concerns or take other action against the country that it believes is not meeting its international obligations.

In the United States, such an approach is quite common. The USTR often issues public notices concerning whether foreign governments have lived up to the obligations outlined in certain of their trade agreements with the U.S. government. If a particular country is found to have failed in this regard, the USTR will often look for remedies, such as having formal negotiations with the party in question or, in some severe cases, even taking retaliatory action against the foreign government, as available under international trade law. In some cases, this sort of action will resolve the issue at question. However, such actions do not always result in success.

Other types of formal government actions, such as meeting on mutual issues of concern, may be brought about through lobbying by industry. For example, service providers that are facing interconnection problems in a particular country may meet with its government and convince the government of the egregiousness of the action of the other country. This may lead the host government to establish a formal dialog on this issue with the foreign government with a focus toward reaching a satisfactory resolution of the issue.

Informal government intervention is another avenue to pursue. Often, companies can convince a government that they are being harmed by the anticompetitive conduct of a foreign service provider or because of the policies of a particular government. In many of these cases, formal action may not be possible, either because there is no mechanism for such recourse or because it may not be politically saleable. In such cases, the complaining party may be able to convince the government to discuss these issues with the foreign government on an informal basis. These issues may be brought up in the context of discussions totally unrelated to the telecommunications issue or in discussions of related issues. The expression of concern by one government to another may spearhead a movement by the other government to take appropriate action. However, success is often dependent on the other government being persuaded to move in this direction.

Another option is to work directly with the embassy of the country where the service provider is located to obtain intelligence on what is going on in the market and in the government on issues of concern. An important role of most embassies is to promote their countries' businesses in the foreign country. Quite often, an embassy will assist the service provider in finding possible avenues of recourse and discussing the issues informally and formally with members of the host government.

Further, aid programs are another method of obtaining liberalization and market access results. Through such a partnership, key goals can be achieved, including improving economic growth, providing greater access to medical expertise and education, and opening up greater access to information to the public at large, among others. Some recent programs have included a USAID initiative in Guinea, which became the first country in

Africa to extend high-speed Internet access to its secondary cities through high-speed satellite and emerging wireless technologies. In Egypt, the U.S. embassy has run workshops on processing online payments to help expand the use of e-commerce. Such programs demonstrate the importance of telecommunications and how opening the markets for these services will result in societal benefits. This often provides a basis for governments to become more receptive to the need to create a competitive telecommunications market.

5.6 The role of competition and international trade law

Another important avenue for exploration is the competition and international trade laws that are in place, both on an international and country basis. These include the antidumping rule imposed by the World Trade Organization and individual nations, international and individual countries' competition laws, and bilateral free trade agreements.

Antidumping essentially involves the imposition of penalties against foreign goods that are being sold cheaper locally than in the country of manufacture, or below the cost of production, if domestic producers can show that they are being harmed by the "cheap" imports. Antidumping regulations are permitted under the world trade rules administered by the World Trade Organization. While competitive pricing is usually of benefit to consumers, in the case of antidumping the complaint lies against predatory pricing, designed to undercut other competitors and drive them out of business. Although this type of action is not directly related to telecommunications services, it can directly impact the industry in terms of impacting the availability of CPE and telecommunications network equipment.

Another option to seeking redress under antidumping laws (and one that is available for both services and products) is the ability to seek a remedy through national and international competition laws. Under this structure, aggrieved parties are able to seek redress against predatory pricing or other similar behaviors under competition law. An example of this is the United States Section 301 of the 1974 Trade Act [20]. Under Section 301 the United States may impose trade sanctions against foreign countries that maintain acts, policies, and practices that violate or deny U.S. rights or benefits under trade agreements or are unjustifiable, unreasonable, or discriminatory and burden or restrict U.S. commerce. In addition, there is a "Telecommunications 301." Under this provision the USTR reviews all trade agreements involving telecommunications products or services to determine whether the foreign country is in compliance with the terms of the agreement, or otherwise denies within the context of the agreement mutually advantageous market opportunities to U.S. telecommunications products and services. Similarly, of the over 130 free trade agreements in place, all of them address competition aspects of trade relations.[2]

5.7 The need for a global regulator

An issue that is often raised in the international arena is whether there is a need for an international regulator to establish a global regulatory framework. This approach has both critics and advocates. Advocates in favor of such an approach point to the success of the European Union in establishing effective competition across their member states. Most critics argue that the effectiveness of such an organization would be limited, especially in light of its inability to obtain jurisdiction over countries because of their national sovereignty. This has been a major hurdle that the EU has had to meet on a constant basis in having member states implement its directives. Other nations, including powerful countries such as the United States, fear ceding the amount of control to such an organization that it would require to be an effective regulatory forum.

It is unlikely that a true global regulator will be established in the upcoming years. Most countries remain hesitant to surrender or turn over areas of responsibility for their economies that they deem falling within their national sovereignty. Nonetheless, there does seem to be a movement afoot to craft regulatory policies that can cut across national borders and, at least in broad concept, be applied in most global markets. This will likely increase as more and more markets open to competition. Accordingly, over the long term, there may be an increasing place for international organizations to step in as policy setters, with national governments retaining more of an implementation and enforcement role. Accordingly, many of the forums that have been discussed in this chapter are likely to have growing importance in the next decade.

---

2. For a discussion of recent anticompetitive conduct by foreign countries as identified by the U.S. government, see http://www.ustr.gov/html/2001-contents.html.

References

[1] http://www.itu.int.

[2] http://www.itu.members/index.html.

[3] http://www.itu.int-t.

[4] http://www.itu.int/itu-d.

[5] http://www.itu.int/osg/spu/intset/index.html.

[6] http://www.itu.int/osg/wtpf/index.html.

[7] http://www.itu.int/itu-t/studygroups/.

[8] Manner, Jennifer A., "The 2000 World Radiocommunication Conference and Third-Generation Mobile Services: The Creation of a New Dynamic," *Catholic University Law School Communications Law Journal,* 2001.

[9] http://www.wto.org/english/news-e/pres97-e/summary.htm.

[10] http://www.ustr.gov/enforcemnt/dispute.html.

[11] USTR Press Release, "USTR to Request Consultations with Mexico Regarding Telecom Trade Barriers," July 29, 2000, available at http://www.ustr.gov/releases/2000/07/0057.html.

[12] OECD, "About OECD: What Is OECD?" 2001, available at http://www.oecd.org/about/general/index.htm.

[13] http://www.infodev.org.

[14] http://www.citel.oas.org.

[15] http://www.eto.dk.

[16] http://www.afdb.org.

[17] http://www.trasa.org.

[18] http://www.apecsec.org.sg.

[19] 19 U.S.C. § 3301 et seq.

[20] 19 U.S.C. § 2411 et seq.

6

Creating a basic corporate strategy for market access

This chapter addresses the first step in the evaluation of global telecommunications markets for investment and entry by a potential service provider—the development of a basic corporate strategy—and provides a case study for further analysis. This chapter should be considered in conjunction with Chapters 7 and 8, which focus on the due diligence process and the actual mechanism and structure available to investors for market entry. The case study addressed in this chapter is further developed at the end of Chapters 7 and 8.

Before delving too deeply into market evaluation, as discussed below, a basic premise must be established: most, if not all, telecommunications service providers have as their prime motive for entering a market corporate profitability. How this profitability is defined is fluid, depending on the overall corporate strategy and goals of the particular entrant. However, as the next two chapters evolve, it is imperative that this principle remains a focus, because it is a key motivator behind almost all of the behavior of entrants into the telecommunications sector. At the beginning of the entry process, the entrant must believe that it will be profitable either directly or indirectly through the market entry. If the entrant does not have this belief, it would not be rational for it to enter the market. Further, the entrant will almost always ultimately exit the market if profit expectations do not near fruition. Accordingly, profitability is a key factor in all steps of creating and executing a market access strategy.

In terms of the actual evaluation of a telecommunications market for entry, most companies establish a general corporate strategy on how they generally plan to enter new markets and expand operations in existing markets. In addition, the corporate strategy also identifies the factors and their respective weight that the corporation identifies as important to make an investment opportunity attractive. Financial considerations generally include a baseline for costs, revenue, and profitability and an associated timeline for achieving a set rate of return. Direct profitability, however, is not always the sole or determining factor in entering a specific market. In some cases, a corporation may adopt an expansion goal for certain markets or product services to attract new customers or retain existing ones that require networks with greater geographic reach or capabilities. However, in general, the net result of the overall entry strategy must be profitable. This has become even a greater focus with the recent turbulent nature of the telecommunications industry.

As discussed below, evaluating a market opportunity and forming a corporate strategy requires a company to identify its products, the technologies it is willing to utilize to provide the product (e.g., wireline, terrestrial wireless, and satellite), and the customer base it desires to serve. Another important consideration, especially in regard to terrestrial wireline basic services, is whether to invest in an incumbent or enter as a new entrant. These and other factors are merged together to formulate a basic corporate strategy, which is utilized to evaluate the attractiveness of individual market opportunities and how they fit into the overall corporate structure and plans. This chapter addresses the key factors that must be inputted into creating the corporate strategy.

Once this basic corporate strategy is identified, the potential entrant is then in a position to evaluate the individual market opportunities that present themselves. Each opportunity requires a due diligence process of sorts, where many key financial, political, regulatory, and technical factors are examined by the entrant. These criteria include, but are not limited to, foreign investment limits; regulatory authorization procedures; local presence requirements; an examination of the legal, political, and regulatory structure; practical impediments to market entry; and the cost of entry. In many cases, new entrants retain experts, such as lawyers and engineering and economic consultants, to assist in this evaluation process. This process is fleshed out in Chapter 7.

Based on the results of the due diligence process, the new entrant must determine if the proposed opportunity makes sense to proceed. If it does, then, as discussed in Chapter 8, the new entrant must determine the form in which it wishes to enter the market. Accordingly, Chapters 7 and 8 build on the next steps the entrant must take and continue the discussion with the case study utilized in this chapter.

6.1 Profitability as a goal

As discussed, a key goal for most, if not all, companies seeking to provide telecommunications services is profitability. Almost every telecommunications company is focused on achieving a certain rate of return on its investments in order to satisfy shareholders or other investors. This rate of return can range from a nominal amount, such as 5%, to more substantial rates of 50% to 60%. Accordingly, a general guiding principle of creating a basic corporate strategy for market access is to ensure that overall the opportunities that are pursued are profitable.

Despite this overarching principle, as discussed below, some companies may choose to enter a market that may not meet their direct profitability goals in order to obtain coverage in an area that provides them with strategic value. Other companies may enter a market where it takes years to make in-roads, facing several years of high losses in anticipation of achieving higher profitability at a later date. Such decisions are very market and corporate strategy dependent. Drivers such as the customer base, what competitors are doing, future market projections, and similar factors are important in evaluating whether a corporation may take advantage of a certain market opportunity. These factors are very market and/or product specific and must be considered by the entrant in the individual opportunity business case. For example, a regional service provider, with a focus on Central America, may find it necessary to enter a market where it does not currently have coverage in order to be more attractive to customers by having a broad geographic reach. Accordingly, even if such a market is not financially attractive, the entrant may find an overriding strategic reason for the entry by being able to attract a broader customer base.

6.2 Formulating a basic corporate entry strategy

As discussed, it is imperative that a telecommunications service provider generally have an overall basic corporate strategy for market entry. An in-place basic corporate strategy ensures that the provider is able to react to potential market opportunities and, where appropriate, work with relevant governments and other key players to create opportunities that it can leverage in the future. In addition, the strategy allows the corporation to structure itself internally in a manner that allows it to operate most efficiently to capture the synergies of its activities. In this regard, the following discussion addresses the key decisions an entrant must make in regard to market entry in order to structure the most appropriate overall entry strategy for its corporate culture, philosophy, and goals.

6.2.1 Identifying a customer base

Perhaps the most important basic strategy decision, and the one that drives almost all others, is the customer base that the entrant wishes to focus on. This decision is integral to evaluating all market entry evaluations. In general, telecommunications customer bases can be broken down into the following categories: business, residential, rural, and high-end user. These categories often overlap, so a business strategy that focuses on one or two customer bases may allow the entrant to also serve customers in the other categories.

Generally, customers who can afford to pay for their services, such as high-end residential users and business customers, are seen as more attractive to most larger entrants and entrants that seek quicker profitability. This is because these users tend to pay their bills promptly and have high demands for telecommunications services. Further, these customers tend to be located in more urban areas, where infrastructure is generally cheaper to build and/or where an existing infrastructure may already exist that can be leveraged.

Because of the inability of many residential users, especially rural and remote users, to pay bills, and because of the widespread geographic dispersion of users, more and more tenders for basic services are seeing less corporate interest. This is in large part because of the low profitability of such services. For example, over the past couple of years several large privatizations and tenders for basic telephony licenses, including in Ecuador and Poland, have failed to reach fruition or reach finalization on a timely basis because of lack of interest on the part of the investment and service provider community.

6.2.2 Identifying a service product

Another important step in determining a basic corporate entry strategy is the determination of the service product or products that the company wants to provide. A few decades ago, such a choice was not as difficult as it is today, because of the smaller amounts of product services available. At that time, services consisted primarily of residential voice, corporate voice, and narrowband data. Hence, basic voice or simple data services (such as facsimile and telex) were the primary services available, and almost all service providers provided these services to their customers.

Today, however—largely because of the influx of new telecommunications service providers, all seeking a market share—technology has blossomed. In the 2000s, the range of products that a company might offer is almost limitless and cuts across traditional service lines. Over time, it is also becoming increasingly difficult to classify service products. Traditionally, products could be grouped as either data or voice products. However, as technology is enhanced, these traditional distinctions become more and more blurred. A good example of this blurring is the concept of voice over Internet protocol. By providing voice services over a traditional data transmission facility or network, service providers have been able to bypass the incumbent service providers in many countries. By bypassing the incumbent, the service provider is able to avoid the large fees that incumbent service providers generally charge for access to their networks. This results in the ability of the voice-over-IP service provider to offer voice services at lower rates than it would otherwise be able to do.[1] Examples of the range of services are depicted in Table 6.1.

Despite this plethora of service products, a corporation must make product decisions or it will be spread too thin in terms of technical resources, personnel with adequate expertise, and finances. Accordingly, most companies focus on certain limited types of services, where they can best fulfill corporate expectations and serve their targeted customer base. This trend has become even more apparent as the telecommunications market's profitability started to decline in the early 2000s and telecommunications service providers were forced to reevaluate their strategies and streamline their operations [1].

1. Potential services can include voice telephony (domestic, local, or international), voice over Internet protocol, Internet services, dial-up services, Frame Relay and other high-speed data services, broadband services (including local and multipoint multichannel distribution services), and even more.

TABLE 6.1 Examples of Telecommunications Services

Local Voice	FRS
Long-Distance Voice	Packet Switching
High-Speed Data	Mobile Telephony
ATM	Mobile Data
Private Line	Point-to-Point Radio
Internet	Air-to-Ground Services
Voice over Internet	

A good example of such a narrowing of focus is WorldCom, Inc. WorldCom is the second largest long-distance service provider in the United States and provides data and other services globally. Its product portfolio includes, as detailed in Table 6.2, a full range of products, from voice telephony, Internet services, high-speed data, paging, and reselling mobile voice wireless services, among others. However, as the economy began to face a downturn in the early 2000s, WorldCom revisited its corporate strategy and began to focus the bulk of its expansion resources on the high-speed data services market with a focus on corporate users using IP technology. In order to do this, it separated its residential long-distance business from its higher-return data service business. It is the former market that has provided it with the largest margins and where it sees continued growth opportunities.

6.2.3 Identifying technologies

Another important basic strategy decision that must be made is the types of broad technologies and delivery methods that the corporation wants to utilize and incorporate for the provision of services. Technologies and delivery methods can include terrestrial wireless, satellite, landline fiber, microwave, and other evolving delivery methods.

TABLE 6.2 Examples of Key WorldCom Data Services

IP VPN	Web Centers
FRS	Internet Access
ATM	IP Communications
GDL	IP Security Services
Web Hosting	

In general, most corporations first decide whether the service product that the corporation would like to provide is technologically able to be carried over wireless or wired infrastructure or a combination of both. Many services may be most suited, from a technology standpoint, to being offered over a single medium or limited types of technology. For example, many providers of high-speed data services are hesitant to offer these product services over the current generation of satellite facilities, because of reliability standards, but these services are able to be carried over a more high-performance microwave service.

In terms of a wireless versus a wireline network,[2] the service provider should carefully consider the following general considerations.

Wireless

- Costs of the regulatory authorization—often spectrum is auctioned for use, generally increases costs of service provision;
- Decreased cost of infrastructure;
- More quickly deployed;
- More subject to environmental issues and weather, such as rain, ice, and snow;
- Generally has lower service availability and reliability;
- Requires more extensive customer premises equipment;
- The degree of regulation. Since wireless services are more recent, there may be less-burdensome regulations associated with providing these services.

Wireline

- Requires an extensive infrastructure;
- Higher service availability and reliability;
- Able to provide higher service-level arrangements and greater reliability;
- Generally less expensive than wireless;
- Requires less customer premises equipment than wireless.

2. Does not discuss satellite communications, which raise additional complex issues.

Additional factors must be weighed, as discussed below, for each individual service product and for the entry.[3] Nonetheless, in developing a new overall corporate strategy, because of the costs and implications of each technological choice, it is imperative that the general technological parameters be established early on, since these will impact major network and entry decisions.

6.3 New entrant or incumbent service provider

The decision on whether to enter as a new entrant or incumbent service provider is often market specific. Further, for many types of services, such as value added or mobile voice, there is only the choice to enter as a new entrant. However, many companies, because of corporate culture and beliefs, establish fairly stringent general guidelines on this issue.

In order to evaluate which is the better approach for an individual company, it is helpful to outline the pros and cons of each approach. The following text outlines typical considerations associated with both types of entry.

6.3.1 Entry as an incumbent

Pros

- Infrastructure in place;
- Large existing customer base;
- Established relationship with the government;
- Experienced personnel;
- Generally has access to scarce resources, such as spectrum and real property;
- Often has substantial cash reserves.

---

3. Other important technological issues include the ability to obtain redundant or back-up facilities. For example, many service providers have been hesitant to enter the South African international value-added services market because there exists only one fiber-optic international link. In most cases, redundant facilities are required for these types of services to ensure that the service has the high reliability that customers seek. Accordingly, for example, corporate technology strategy regarding these types of redundant facilities may have to be waived in order to enter some markets where standards are less stringent.

In order to capitalize on new opportunities, the entrant needs to stay abreast of and anticipate key country and regional developments in the areas of privatization, liberalization, and the creation of an effective regulatory regime. Particular emphasis should be placed on monitoring the product services that the company has identified as integral to its basic strategy.

In addition, the entity must pursue certain key activities if it is to position itself as a player. Specifically, it is important that the entrant develops and maintains a calendar of key events for the market under consideration. In addition, the entrant must identify opportunities to influence processes alone, with other parties, or through government channels.

If an opportunity does warrant such additional examination, a team of experts will be pulled together to examine individual opportunities through the due diligence process, as discussed in Chapter 7. Subsequent to this process, if an affirmative decision is made to proceed, the entrant will form a team of experts to structure and execute the deal.

6.6 Case study: MarathonTel, an overview of the company and its corporate strategy

The following text is a factual overview of the case study to be utilized throughout Chapters 6, 7, and 8. The first part of the case study, set forth below, discusses the subject of a corporation, MarathonTel, and its basic corporate strategy for entering new markets.

MarathonTel is a U.S. company and has been a provider of telecommunications services since the early 1980s. It has primarily focused on traditional voice telephony in the U.S. long-distance market, although it has begun to explore potential entry into the U.S. local service market. In addition, it has a very profitable business in providing corporate communications services. In this regard, MarathonTel has expanded its network around the globe to provide high-speed data services, such as FRS and ATM.

Because a large percentage of MarathonTel's clients are active in Eastern Europe, this has been an area of focus for expansion of its network. Accordingly, to date it has entered the Czech Republic, Romania, Hungary, Bulgaria, Russia, and Ukraine. MarathonTel would like to continue this geographic expansion and has identified other key Eastern European markets that may be of interest. In general, MarathonTel believes that a high rate of return is imperative to make a business opportunity attractive.

There have been exceptions to this but only based on important strategic considerations, such as a large number of customers having requirements in that particular geographic region. Overall, it seeks to obtain at least a 25% rate of return on its investment within 2 years of operations.

Establishing its international operations has generally been fairly formalistic for MarathonTel. MarathonTel looks to place a simple data switch in a country over which it will operate its data services. In order to provide its services, it also needs to obtain local and international capacity from authorized service providers. In terms of technical support of its switch, most can be handled remotely, with MarathonTel contracting with a local entity for technical support. To date, although MarathonTel's international services are technically capable of being provided over wireless technology, its engineers have been hesitant to utilize this technology because of the impact of certain environmental factors, most notably rain, upon its performance levels and availability.

In terms of its corporate strategy, MarathonTel has focused on traditional services in the U.S. market. However, it has limited itself to corporate services around the globe. Because of MarathonTel's decision to provide international high-end services, factors such as reliability, availability, and the time to repair are imperative. Accordingly, MarathonTel has determined that it will utilize primarily fiber optics for both the domestic and international portions of its communications. In rare cases, for the domestic portion, MarathonTel will utilize microwave as a fill-in solution, but only if certain service levels can be met.

In terms of MarathonTel, because of its historic competitive bent, it has primarily focused on new-entrant opportunities. This also gives MarathonTel the ability to enter the market at a lower initial cost, since it is not purchasing infrastructure. In addition, it believes that it is best to establish a regional network in Eastern Europe to serve the general range of customers located there. MarathonTel has researched the regional and global market and found a void of service providers that offer good coverage in Eastern Europe. To date, it appears that most service providers in the region have only accessed a couple of key markets and no player is truly regional.

References

[1] Norris, Floyd, "It's Not Just AT&T—How Telecom Became a Black Hole," *New York Times*, February 16, 2001, available at http://www.newyorktimes.com.

[2] Tagliabue, John, "Telefonica's Brave New World: A Strategy with a Latin American Centerpiece," *New York Times*, May 13, 2000, available at http://www.newyorktimes.com; and "Spanish Phone Utility Extends Its Latin American Leadership," *New York Times*, August 4, 1998, available at http://www.newyorktimes.com; http://www.telefonica.com.

[3] http://www.globalcrossing.com.

7

Evaluating individual market opportunities

7.1 Overview

Chapter 6 focused on developing a basic corporate strategy for market entry. This chapter addresses one of the key next steps in entering a new market: the evaluation of the market opportunity. Once the basic corporate strategy is identified, the potential entrant is then in a position to evaluate the individual market opportunities that present themselves and how to structure its market entry. Although these steps often occur simultaneously, for purposes of clarity the evaluation of the opportunity is examined in this chapter and the structure of the entry is addressed in Chapter 8. In addition, the review process of the evaluation is briefly addressed in this chapter and expanded on in Chapter 8.

Each new market entry opportunity that is of interest requires a potential entrant to engage in an extensive analysis, including a due diligence and an examination of the many key financial, political, regulatory, and technical factors. As discussed below, the analysis first focuses on the three key factors contained in the basic regulatory formula—privatization, liberalization, and effective regulation—and also includes an analysis on a deeper level through the due diligence process.

Specifically, the due diligence process is concerned with examining the actual criteria that make up the basic regulatory formula and other relevant benchmarks, such as the state of the economy of the target market. These criteria include, but are not limited to, foreign investment limits; regulatory authorization procedures; local presence requirements; an examination of the legal, political, and regulatory structure; practical impediments to market entry; and the cost of entry. In many cases, new entrants retain experts, such as lawyers and engineering and economic consultants, to assist in this evaluation process. This process can be quite resource and time intensive, with site visits, meetings with key government agencies, the creation of advocacy strategies and documents, participation in legal and regulatory proceedings, and other required actions.

Simultaneously with the due diligence, the new entrant also must determine the form and structure with which it wishes to enter the market, since quite often the potential form of entry also influences the criteria that are looked at in the due diligence process. However, for purposes of clarity the due diligence process is addressed in this chapter, and Chapter 8 focuses on the structure of the entry.

7.1.1 Identifying opportunities

Before outlining the actual due diligence process, it is important to understand how a potential entrant is able to identify opportunities of interest that may fit in with the entrant's basic corporate strategy. Essentially, there are three main types of opportunities:

1. Opportunities that are presented to the potential entrant;
2. Opportunities that the potential entrant seeks out;
3. Opportunities that the potential entrant creates.[1]

The following text is a short discussion of each of these opportunity types.

1. It is important to remember that no matter what type of entry is pursued, the entrant may find it necessary to seek changes to the legal, policy, or regulatory structure in order to make entry more attractive. This process may be resource intensive and time consuming, depending on the types of changes being sought. The changes that are needed are in part what the due diligence process will reveal. Therefore, the nature of these changes may directly impact the attractiveness of the opportunity.

7.1.2 Opportunities presented to the potential entrant

There are several situations that exist whereby market entry opportunities may be presented to the potential new entrant. The first involves governments that are seeking to open their telecommunications market or specific sectors within this market to new investment, either through the issuance of new authorizations (generally for basic services) or the privatization process. In such situations, the government of a country will often retain experts from investment banks or consulting companies, such as accounting firms, and go on a "road show" in an attempt to interest potential licensees or bidders.

A good example involves Singapore and its decision to fully open its market in 2000 for basic services, 2 years earlier than originally planned [1]. In this situation, the government sent a team of representatives from various government agencies, including the regulator, to the United States and other countries to meet with companies that might be willing to pursue basic service authorizations. At these meetings, the Singapore government representatives were able to fully brief these new entrants about the market potentials, and the potential entrants were given the opportunity to voice their concerns and interests in private. This process helped to boost investor confidence and interest. In addition, by laying the groundwork ahead of time, the government was certain that there would be interest and support when it made an official announcement to open its market to new service providers. The Singapore government was right and within a couple of years there has been the creation of vibrant competition operating in the Singapore telecommunications market.

Another common situation where opportunities seek the potential entrant out is where a consortium or joint venture is being put together to seek market entry in either one or more markets. In this situation, companies will seek to align themselves with other entities that can assist with their entry into the new market or markets. An example of this is when the Israeli market for long-distance services was being opened to limited competition in the early 1990s [2]. In this situation, because of the need for expertise and capital, combined with the requirement for partial Israeli ownership, many companies put together consortia to bid for the authorizations, including Barak[2] and Golden Lines, the two entities that success-

2. For example, Barak's foreign investors include Sprint, France Telecom, and Deutsche Telekom. "The World's Strongest Telecommunications Alliance: About Barak," available at http://www.barak-online.net/html/about2eng.asp.

fully obtained regulatory authorizations. Often, these consortia start with just one or two members of the consortia and then branch out to include other investors that might be a good strategic fit either through their operational abilities, finances, or local presence.

7.1.3 Opportunities that the potential entrant seeks out

Many companies have tasked employees or outside consultants to monitor countries of potential interest for entry to determine when potential new opportunities arise. This requires the entrant to monitor policy and legal changes to see if potential entry into the market in a manner that the entrant feels is meaningful becomes available.

Accordingly, once a new entrant opportunity is identified, generally using the basic corporate strategy as the flagging mechanism, this opportunity will be raised to the attention of the appropriate business unit. The business unit will often ascertain, with input from other corporate organizations, if the opportunity is of interest for the company to pursue. An example of such an approach is a company that is trying to obtain Internet service authorizations in all Latin American countries as part of a regional strategy. This company may have its in-house lawyers monitor legal changes in key countries so that the business development team can be alerted to when the opportunity arises to obtain such an authorization. The business team will then evaluate the company's interest in pursuing the opportunity on a high level before proceeding with the more extensive due diligence process, generally involving the marketing, engineering, and financial divisions of the company.

7.1.4 Opportunities that the potential entrant creates

The final form of opportunity is that which the potential entrant creates. This form of entry is generally more resource intensive then the other forms and is often harder to achieve. In this situation, the entrant may determine that it wants to enter certain markets to provide certain products through a set entry mechanism. In order to achieve this plan, the potential entrant may need to seek and encourage extensive changes to the country's legal, regulatory, and/or policy regimes.

A good example of a company that had such an ambitious plan is Global Crossing. Global Crossing's strategy for market entry can be characterized as follows:

1. Determining which markets it wanted to enter;
2. Seeking authorization to build wholly owned international cable capacity to that location;
3. Obtaining cable landing authorizations in that market;
4. Obtaining a full facilities-based authorization, allowing it to provide service using its own facilities, including international cable capacity.

Since such an entry approach was not authorized in several countries that Global Crossing identified, it expended numerous resources in order to seek the policy, regulatory, and legal changes that it would allow it to fulfill its corporate strategy. Specifically, in markets such as Argentina, Global Crossing expended large amounts of resources to lobby the government to make the sort of regulatory changes that would allow it to enter the market consistent with its business plan and in accordance with its proposed strategy. All total, Global Crossing was successful in over half of the markets it initially identified [3].

7.2 The due diligence

Once an opportunity is presented to the potential new entrant, it must begin the process of determining whether it is an opportunity that is rational to pursue. In order to do so, it must determine first if the opportunity is consistent with the company's corporate strategy. If it is not, then the company needs to ascertain whether it is an opportunity for which it wishes to make an exception to its overarching philosophy. Such an analysis must include an examination of whether the company has or can acquire the resources and skills to take advantage of the opportunity, such as the required funding. If an affirmative decision is made to proceed, the basic corporate strategy of the company may be revised in the near future to reflect this new area of business. Second, the entrant generally examines the factors addressed by the basic regulatory formula in the market in which the opportunity is presented. Third, the entrant begins the due diligence process. Based on these reviews, the company is then able to begin a more in-depth analysis of the opportunity. This review process is discussed in Section 7.2.3.

7.2.1 The overarching review of the basic regulatory formula factors

As discussed, the goal of most new entrants is to enter a market in which a level playing field has been created, in which competition can blossom, and where the entrant can ultimately compete successfully. Accordingly, new entrants often use the factors contained in the basic regulatory formula as a first-level review to see if the market they are entering is competitive. The general questions that the entrant would ask are outlined below. Of course, the specific questions would be tailored to the individual entry plans of the service provider.

1. Has the government privatized or does the government have plans to privatize the incumbent service provider or providers? Is the privatization complete or partial? If partial, what is the time frame for complete privatization? How has the privatized service provider fared in the market?
2. Has the government liberalized the telecommunications services market and, if so, to what extent? If not, are there any plans to liberalize the market? Has the liberalization process been successful?
3. Does effective regulation exist in the market and where is the regulatory structure heading?
4. Does the government have a rational plan for deregulation?

The broad answers to these questions will color the due diligence analysis. For example, if the opportunity is to engage in the privatization process and no competition exists in the market, most of the questions about the competitive market, such as dominant service provider safeguards and interconnection, will not be addressed. Conversely, the due diligence process will often become more complex when it involves a market that is just starting to foster competition. In addition, each answer will be an integral part of the analysis that the new entrant relies on in determining whether an entry opportunity is attractive to pursue.

7.2.2 The due diligence checklist

In evaluating a market for entry, different telecommunications companies and different types of telecommunications service providers have different requirements. However, in most cases, in order to evaluate the opportunity, the entrant will create a due diligence checklist. This checklist is relied on to evaluate the specific market opportunity and often includes many of the general items discussed below. As seen in the case study on MarathonTel at the end of this chapter, the due diligence list is always adapted to the specific market entry opportunity at hand, including the specific service product and customer base.

The performance of the due diligence process requires the entrant to assemble all relevant documentation, including laws, regulations, economic information, forecasts, and the competitive information available. In addition, the scope of the opportunity presented must be identified and the due diligence process must be crafted around the scope. Next, a regulatory and opportunity checklist linked to the opportunity must be developed, with assignments divided among key company personnel with whom specific expertise lies. In addition, during the due diligence process it is imperative that the calendar of events must be monitored even more closely. As the opportunity is further mapped out, a strategy for influencing the process from a business, technical, economic, and regulatory perspective must be established. As discussed, all of this information is analyzed together and looked at in conjunction with the business case.

The following list presents a broad, general view of the issues that are often examined during the due diligence process, with a brief explanation of what each item means.

- *WTO membership:* Is the country a member of the WTO and has it introduced an offer to the Basic Agreement on Telecommunications? If so, what does the offer provide? Has the government lived up to the terms of the offer and the Regulatory Reference Paper? If the government is not a member of the WTO or has not provided a WTO commitment, what is the time schedule, if any, for membership and/or submission of a commitment?

- *Foreign investment limits:* Do direct or indirect limitations exist on the investment in telecommunications service providers by foreigners for the relevant service? Are these limitations full or partial? Do these limitations also impact the ability of the new entrant to have operational control of the service provider?

- *Local presence requirements:* Is a local presence required in order to receive a registration or authorization to provide telecommunications services? Do these requirements require a specific form of entity, such as a corporation or partnership? What must the charter of the local entity provide for? What are the obligations the entity will have to meet (e.g., human resources, technical, tax, and so on)?

- *Cost of entry, both regulator fees and other government fees and taxes:* What fees must be paid for the initial authorization and what subsequent regulatory fees, including renewal fees, must be paid? What taxes is the service provider subject to? Are import fees applicable on equipment imported to provide services? What other financial liabilities will the service provider find itself subject to? Are there methods to minimize the impact of such fees and taxes? How do these fees impact the business case?
- *Licensing authority:* What is the authority for the initial authorization and what is the process utilized? Is the process fair and transparent? What is the time frame for authorizations being issued? Is there a process for issuing new services? Can interested parties provide input into the authorization process?
- *Regulatory process:* Is the regulatory process fair, nondiscriminatory, and transparent?
- *Dominant service provider safeguards:* Does the regulatory regime provide for dominant service provider safeguards? If so, what are these and have they been enforced and effective?
- *Resolution of disputes between authorized service providers and the regulatory authority:* What is the means of dispute resolution between the regulator and service providers? Is there recourse to a neutral third-party arbitrator or recourse to the courts? What are the steps necessary to be able to take advantage of the dispute mechanism?
- *Performance guarantees and bonds:* Are specific build-out or service guarantees required to be given by the service provider? Are authorized service providers required to post bonds to ensure that they meet these obligations? If the service provider fails to meet these performance levels, what are the penalties, if any?
- *Terms of forfeiture:* What are the terms of the regulatory authorization forfeiture? Is the process fair and transparent? If an authorization is forfeited, is there an appeals process and can the entity receive additional future authorizations?
- *Sanctions and penalties:* What behavior by an authorized service provider is subject to sanctions and penalties and what are these?

Are there sanctions and penalties imposed on the unauthorized provision of services? Are these sanctions and penalties imposed where warranted and in a fair manner?

- *Authorization modification, assignment, and termination:* When will the regulatory authorization be modified, assigned, and terminated? May interested parties participate in this modification and assignment process?
- *Tariffs and rates:* May the service provider set its own rates and when will the regulator review these rates? When will the incumbent and its customers be constrained by rate regulation? What types of tariff obligations exist for what types of services provided? Has rate rebalancing occurred in the market?
- *Competitive and monopoly services, periods, and territory:* What services are authorized to operate on a competitive basis? What services are still under monopoly or limited provision and what is the geographic territory covered by the monopoly? What is the time frame for opening these services to competition?
- *Rights of way and eminent domain:* What is the ability to obtain access to real property and easements and the process and time frame for doing so? Are the same rights accorded to both incumbent service providers and new entrants in regard to the terms and conditions for access?
- *Spectrum availability:* What is the ability to obtain spectrum? How is spectrum awarded? Is it through a competitive bidding process or other mechanism? Is spectrum allowed to be utilized as an adjunct to terrestrial landline services? Are there spectrum caps? This examination may be expanded on if the service is to be primarily provided through a wireless technology, such as satellite, as depicted in Figure 7.1.
- *Participation in numbering plan and number portability:* To the extent basic service providers are involved, how is the numbering resource structured and allocated? What is the participation of the private sector in the process?
- *Interconnection rights and responsibilities:* What is the ability to obtain interconnection, if the entrant is new, or the terms and con-

Figure 7.1 Satellite technology.

ditions interconnection is supplied under, if the entrant is an incumbent? What happens if an incumbent service provider refuses to provide interconnection on a fair, cost-based, and nondiscriminatory basis? What recourse is available at the regulatory and judicial levels? Is there a regime in place to address dispute settlement? Will such disputes be resolved on a timely basis?

- *Network modernization, expansion, and development expansion:* What commitment must the authorized entity meet in terms of build-out and expansion of its network?

- *Obligations in emergency situations and national security issues:* What national security and other public interest requirements have to be met? What are these requirements? Will the service provider be compensated for complying with these obligations?
- *Billing and collection procedures:* What mechanisms are available to the service provider to collect bills? What has the payment history of customers been?
- *Service interruptions:* What sort of requirements does the provider need to follow in case of a service interruption, whether voluntary or involuntary, including crediting customers?
- *Service quality standards:* What sort of service requirements must the provider meet? For example, what sort of reliability and ability to meet mean times to repair during service outages are covered?
- *Reporting information, inspection, and audit requirements:* What are the reporting requirements the new entrant must meet? Is confidentiality afforded to these reports? What types of infrastructure and audit requirements are imposed? If there are audits and inspection requirements, what is the trigger for these to occur? What happens to the results of these?
- *Regulatory authorities involved—their capabilities and their relationship to other government agencies:* Is the regulator an independent decision maker? Is the regulator well positioned to be able to work well with other government agencies that may impact its operations and its subject matter? What is the expertise of the personnel of the regulator and its size? Is the regulator well funded and from where does the funding come? What other government entities may have jurisdiction over the service provider and in what manner?
- *Enforcement of concessions:* Are the regulator and other involved agencies active enforcers or are the regulations being bypassed illegally?
- *Service complaint processing:* What sort of attention are complaints given and how are they connected into the enforcement process?
- *International participation and agreement authorization:* Is there a role for service providers in the international telecommunications arena, or are service providers excluded? For example, may the service

providers provide input to the government prior to international meetings and may the private sector participate in at least some of these international meetings?

- *Technical standards:* How are technical standards set? Is this through a government or industry endeavor? What mandatory standards are there?
- *Competitive protections and antitrust enforcement:* What role does the antitrust agency play in enforcing competition policy? Is the antitrust agency able to step in to assist enforcing, where necessary, the creation of a level playing field? How has the antitrust agency acted in other similar fields?
- *Ability to obtain capacity (local, national, and international):* As a service provider, how is capacity obtained from other service providers that control infrastructure? Are there rules in place to ensure capacity is provided on a timely, nondiscriminatory, and cost basis? Do all types of capacities fall under the interconnection rules?
- *Consumer protection provisions, accountability, and penalties:* What sort of rules exist to protect consumers from unfair practices and poor service? What sort of penalties are imposed on service providers for failing to treat consumers in a fair manner? What sort of review process of the penalty process is available?
- *The competitive environment:* What sort of competition exists in the market and what is the outlook for increased competition? How have service providers fared in the market?
- *Universal service and rural coverage arrangement:* Is universal service contributed to in a fair and nondiscriminatory manner? What obligations are imposed on service providers? If no universal service policy currently exists, are there plans to implement such a policy?
- *Repatriation of profits:* Can the service provider repatriate profits? What limits are there on the free movement of capital?
- *Political stability:* Is the government politically stable? What are the ground rules for when a new political party or person takes control of the government?

- *Economic outlook:* What is the economic outlook for the market? Is the economy suffering from high levels of inflation? What is the strength of the relevant currency? What large companies currently operate in the market?

7.2.3 Introduction to the review process of the due diligence analysis

After the due diligence process is complete, the company may begin to examine in detail the attractiveness of the opportunity and determine its interest in going forward. This involves looking at the information gathered in the due diligence process and analyzing it in conjunction with a financial review of the market and the specific opportunity. In addition, the entrant may also use this analysis to lobby for certain changes in the regulatory or legal regime of the market.

Although a new entrant may choose to enter a market where true competition does not yet exist, for the entrant to be able to survive in such a market it has to be a conscious choice and one in which the entrant is prepared to accept the roadblocks that are likely to arise as it deploys its services and operates in the market.

Of course, there may be situations where the new entrant enters a market where the basic regulatory formula has not been met and where competition is curtailed. This does not mean that a potential entrant will always shun such a market. There may be situations where the new entrant believes it is worth a fight to gain a foothold in the market. For example, in Poland, Netia, the second largest service provider has made a rational decision to enter the market and build its infrastructure, with the realization that the incumbent service provider in Poland may not provide access to essential facilities and its networks on fair terms and conditions. Despite numerous setbacks, Netia has persevered and is likely to become a successful service provider.

There may be other situations where the new entrant feels that it can also exploit the limited competition. A good example of this was SBC's investment in the South African incumbent service provider, Telkom South Africa. By investing in Telkom South Africa, SBC was able to obtain an exclusivity for basic services for several years. During this time, Telkom has been able to take advantage of the lack of competition and improve, expand, and modernize its telecommunications network.

7.3 Continuation of case study: MarathonTel expands its corporate services to Eastern Europe

This section addresses the potential opportunity in an evolving Eastern European market. Following this discussion, the case study walks through the due diligence process that MarathonTel goes through in evaluating this opportunity for the provision of high-speed data services.

7.3.1 The opportunity

The focus is an emerging economy that is growing in terms of its importance as a business hub in Eastern Europe. It has begun to liberalize its market for some telecommunications services, including long-distance and data services. However, local wireline service is still provided by a monopoly service provider, Telecom, which is majority owned by the government. In addition, Telecom still controls the lion's share of the telecommunications market in general. Entry into the newly competitive markets has primarily been by locally owned and operated companies or joint ventures, since up until last year, foreign ownership was restricted to 49% investment and no control. This past year, this law was changed to eliminate all such restrictions. However, since the date the law became effective, elections were held and a more socialist regime was installed. This regime has introduced legislation that again would restrict foreign ownership and control of telecommunications service providers.

The subject country is seeking admission to the European Union. Accordingly, it has begun to bring its regulatory regime into compliance with EU telecommunications regulation. Therefore, its framework laws are fairly consistent with the EU directives, but, unfortunately, the government has failed to enact the necessary regulations that would ensure that Telecom provides interconnections and access to new entrants. The country's accession to the EU is estimated to take place in 5 more years.

The country is also a member of the WTO but provided little in terms of a commitment. It did agree to the Regulatory Reference Paper but only committed to any form of market opening in 6 years and does not appear to be moving in a faster manner than committed.

The country has been identified by MarathonTel's marketing team as its highest priority to enter in the next year for its corporate service providers, because of the proliferation of high-technology firms opening offices in the domestic market. MarathonTel believes it can gain the business of many of these companies for its highly reliable, high-speed data product.

7.3.2 The review process

MarathonTel, in response to its marketing team's demand to enter the new market, has gathered its team of market entry experts. The team is evaluating the overarching regulatory and market opportunities to determine the best form of entry into the market. It has determined that the potential entry into the data service market is consistent with its basic corporate strategy. Accordingly, MarathonTel has determined that it should move forward with the review process of the market and, subsequently, as discussed in Chapter 8, in determining the best structure with which it should enter this market.

7.3.3 Basic regulatory review

Unfortunately, the MarathonTel team of experts believes that the market has far to go to create an effective competitive marketplace. As discussed, in terms of privatization, the incumbent service provider is only partially privatized, and the government still holds a sizable stake. The government has made some strides in its liberalization efforts in that it has opened the markets for data and long-distance services. Nonetheless, the uncertainty concerning foreign ownership restrictions limits the ability to say that even these services are truly open, especially to foreign-owned service providers, such as MarathonTel. Rules on interconnection, the ability to obtain local and international capacity, and dominant service provider safeguards are still being developed.

Further, in terms of creating an effective regulatory regime, the market has a long way to go. Although the basic regulatory structure is in place, MarathonTel, based on its experience with other service providers, believes that the missing regulation and the untested regulatory regime may mean that effective competition is a long way off. In fact, based on the regulators first 5 months of existence, it appears that the regulator will be slow to act on instituting an improved regulatory regime, because of its almost total inaction.

7.3.4 Due diligence review

In terms of its due diligence process, MarathonTel focused only on the issues that are relevant to its provision of corporate services. A large portion of the due diligence process focuses on the practical ability of MarathonTel to obtain the capacity it needs to provide its services internationally and to local customers on a timely, cost-effective, and reliable basis. The due diligence review includes the following:

- *Competitive overview:* The market is still dominated 95% by Telecom, the basic service provider. The other 5% of the market is primarily made up of three other service providers that hold different types of authorizations, including long-distance and data services. Overall, all service providers have had trouble gaining much momentum at entering the market, since there are no regulations governing access to the Telecom network. Similarly, local zoning regulations have made it quite complex for new entrants to erect their own infrastructure. New entrants must obtain complex authorizations, while the incumbent service provider has no similar requirements. Accordingly, most service providers are still de facto required to go through Telecom to obtain infrastructure.
- *Foreign investment limits:* The government has currently not imposed any foreign ownership restriction on the ability of a foreign company to obtain an international data transmission authorization. However, certain pending legislation would restrict this ability. Accordingly, the outcome of this situation is currently uncertain and may negatively impact MarathonTel's ability to hold a wholly owned service authorization in the future.
- *Local presence requirements:* The government has instituted a registration requirement for international data transmission services, which covers the corporate services that MarathonTel wants to provide. As part of this process, MarathonTel would be required to form a local legal entity.
- *Ability to obtain capacity (both local and international):* A survey of the ability of existing service providers to obtain local capacity reveals that Telecom has been quite slow at providing capacity. To date, there are no regulations that require Telecom to provide such capacity. Accordingly, this has been flagged by MarathonTel's assessment team as a potential problem that will need to be addressed.

  In terms of international capacity, the three existing service providers, in addition to Telecom, all have international capacity, which they have built out. Accordingly, unlike in the local capacity market, international capacity is more easily obtained.

- *Cost of entry—both regulator fees and other government fees and taxes:* The registration procedures are fairly simple: A form is filed and becomes effective in 30 days from filing unless an objection is filed. However, hiring local counsel is imperative in order to be sure to meet all the requirements and to have the necessary lobbying effort in place to address any issues that may arise. Specifically, the regulator often requires supplemental information to be filed that is best provided by local counsel. The registration fee is a mere processing fee of the equivalent of $500. In terms of taxes, they are fairly standard, including VAT taxes of approximately 20%.
- *Licensing authority:* The regulatory authority is the Communications Commission (CC). CC is a relatively young and inexperienced regulatory agency. In addition, the legislature, many of whom do not agree with the liberalization process, has only provided the CC with a small operating budget. To date, the CC has been quick on authorizing service providers but has not taken much enforcement action. Further, the CC has been extremely slow to issue any new regulations. MarathonTel's assessment team is worried about the ability of the regulator to act in a an effective manner to ensure that Telecom complies with its regulatory obligations and to institute the stringent regulations that will be required to ensure that Telecom does not abuse its dominant position.
- *Resolution of disputes between authorized service providers and the regulatory authority:* No regulations issued to date. This concerns the MarathonTel team, since it means there is no recourse to a third party if an authorized service provider has a dispute with the regulator.
- *Terms of forfeiture:* Since a simple registration process with almost no obligations imposed on the service provider is all that is involved for data services, the forfeiture process is rarely used. A forfeiture is required when the service provider provides a service that it is not authorized to provide. This is a strict liability system.
- *Sanctions and penalties:* To date, the government has been lax on imposing sanctions and penalties, primarily because there are few that have been enacted. Accordingly, there are some illegal service

providers operating in all sectors. This may directly impact the ability of authorized service providers to meet their target profitability.

- *Authorization modification, assignment, and termination:* Since the registration process for data services is fairly easy to comply with, there is little problem with revising plans. The authorization is valid for 5 years with a presumption of renewability. Transferring regulatory authorizations is a fairly easy process because of the simplicity of the registration process.

- *Tariffs and rates:* Rates that MarathonTel will charge for its corporate services would be unregulated. In terms of the local capacity and the international capacity, there is a tariff requirement, but the rates are not specifically set by the regulator. This is of some concern to MarathonTel, since it may be subject to artificially high rates set by Telecom, especially in the local capacity area, where there is no competition.

- *Competitive and monopoly services, periods, and territory:* Telecom, the incumbent service provider, has a monopoly for domestic services (except data) but still controls a large number of the other telecommunications markets. In addition, as the incumbent service provider and the former monopoly of all services, it has control of almost all infrastructure.

- *Rights of way and eminent domain:* This has been a problematic area for many service providers. Because of the former communistic political regime, numerous farmers have collective rights to the land. Accordingly, to lay fiber, new entrants must obtain approval from numerous private citizens. Unfortunately, since the regulatory regime has not yet eliminated the favored status of the incumbent service provider, Telecom does not have to go through the same cumbersome process.

- *Interconnection rights and responsibilities:* The interconnection regime that has been created is very weak.

- *Billing and collection procedures:* Corporate service providers have very little recourse for unpaid bills. They can ultimately go through a collection process with the courts, but this is long and protracted. Fortunately, most of MarathonTel's customers are based in the United States and will be billed at their U.S.-based headquarters for the MarathonTel corporate services.

- *Service interruptions:* There is no specific regulation on this issue. However, because of the poor and aging infrastructure in the country, service outages are common on the Telecom network. Accordingly, MarathonTel may not be able to meet its typical service-level guarantees it provides to its customers. This may subject it to either a loss of potential customers or offering customers less than its standard level of service quality, impacting the prices it can charge customers.

- *Reporting information, inspection, and audit requirements:* Corporate service providers are required to file annually a list of customers with the regulator. There is no guarantee that this information will be treated confidentially. This is particularly troublesome, because it provides other service providers with the opportunity to obtain this information.

- *Regulator's relationships with other government agencies:* Because the regulator is quite young, it is also very inexperienced. Further, its staff is primarily composed of former Telecom personnel, making it likely to favor Telecom, at least until the market further matures and the regulator is able to hire experienced personnel from other sources.

- *Enforcement and approval authorities:* To date, the approval process has been quite expedient. However, enforcement has been almost nonexistent. Telecom, primarily because it still has not been labeled as a dominant service provider, has been able to engage in anticompetitive conduct without any enforcement. The regulator, however, has been very aggressive in ensuring that unauthorized providers of services are promptly shut down.

- *Hearing and appeals process:* The regulator has the right to hear all appeals from its rulings; however, there is no time schedule for such hearings to be held. Further, the next step in the appeals process is unclear.

- *Competitive protections and antitrust enforcement:* No competitive protections are in place. The antitrust authority has not actively participated in the telecommunications sector. Accordingly, this provides little recourse to authorized services seeking to restrain Telecom should it act in an anticompetitive manner.

- *Consumer protection provisions, accountability, and penalties:* None instituted.
- *Universal service and rural coverage arrangement:* Only imposed on the basic service providers. Accordingly, MarathonTel will not need to budget such costs.
- *Political stability:* This is a fairly stable country. However, it has begun to return to its more socialist roots, as in the last election when the Socialists retook control of the government. Nonetheless, it is unlikely that it will return to being a communist country.
- *Economic outlook:* The 1990s brought unprecedented economic wealth to the country. However, the economy has begun to slow. Nevertheless, many major companies entered the market during the economic boom of the 1990s and it appears that it will remain a vibrant market, especially in the high-technology sector, because of its highly educated workforce.

7.3.5 Review of the opportunity

MarathonTel believes that because of the availability of authorizations, and the upcoming EU membership of the country, it is wise for it to enter the market to serve its customers. However, the inability of other service providers to obtain capacity and the possible imposition of foreign ownership restrictions have made MarathonTel concerned about the form and structure of its entry. Accordingly, MarathonTel will, in Chapter 8, carefully evaluate how it plans to enter this market and how to structure its venture.

In terms of addressing the shortcomings of the due diligence list, MarathonTel plans to institute a heavy lobbying plan. In order to be most effective from both an advocacy and cost perspective, MarathonTel has identified four key issues in which it will at first focus its efforts to have remedied. These issues are:

- A fair, cost-based interconnection regime;
- The imposition of a dominant service provider safeguards;
- The creation of a more independent and effective regulator;
- Ensuring no foreign investment limits are imposed.

To achieve these goals, MarathonTel is retaining a consultant with regulatory, legal, and political expertise to work with the regulator, the executive, and the legislative bodies to create an acceptable regulatory regime. In addition, MarathonTel has discussed its concerns with its potential in-country customers and has encouraged them to also seek these changes within the country. Finally, MarathonTel is evaluating the use of both U.S. government resources and international resources in obtaining a successful outcome to its lobbying efforts.

References

[1] "Singapore Infomap," available at http://www.sg/flavour/profile/pro-transport9.html; and "IDA to Launch Consumer Education Programme on Start of Competition in Public Basic Telecommunications Services," available at http://www.ida.gov.sg/Website/IDAContent.nsf.

[2] "Israel Telecomms Market Open but Competition Absent" (last modified 2001), available at http://news.itspace.com/telecom.

[3] Global Crossing Fact Sheet, available at http://www.globalcrossing.com/presskit.

8

Structuring global telecommunications ventures

8.1 Overview

This chapter supplements the discussions in Chapters 6 and 7 on evaluating new market opportunities. Specifically, this chapter will examine how to structure a global telecommunications venture in a new market through various structural forms of entry and the key considerations that must be taken into account by the entrant in selecting the most suitable form. In addition, the case study concerning MarathonTel will be expanded, based on the earlier corporate strategy and due diligence discussion, focusing on the appropriate form of entry for MarathonTel to pursue in entering the proposed Eastern European market.

As discussed below, there are several major forms of entry into a new market. These include:

- Full operating company;
- Joint venture;
- Partial entry through a deployed network;
- Leveraging a third-party network;
- Reseller agreements.

This chapter will examine each form of entry and the issues that must be considered in regard to each. Each form of entry has certain benefits and disadvantages, which vary depending upon the indirect and direct goals of the entrant for the particular market. Accordingly, selection of the form of entry depends on many factors, including the basic corporate strategy, the due diligence process, and the opportunities that are available in the market.

Further, each form of entry mechanism is not exclusive. Entry through multiple forms is possible and often pursued by a single entrant. In many cases, entry through multiple forms is sequential, based on the maturity and competitive situation existing in the relevant market. For example, when a market is still fairly closed, the only possible form of entry may be through a reseller agreement. As the market begins to liberalize, however, further opportunities for entry may open up, including deployment of a limited network. Upon the introduction of even fuller competition, the entrant may be able to establish a full operating company. During these phases, depending on the needs of the entrant and its customers, the earlier entry mechanisms may be disbanded, consolidated, or continue to be operated separately.

8.2 Overview of the general forms of entry

As discussed, there are five major forms of entry into a telecommunications market. The first two forms, a full operating company and a joint venture, generally involve the most substantial investment and resource commitment on the part of the entrant. The next largest commitment is partial entry through a deployed network. Commitments typically associated with these forms of entry include funds for limited network and limited personnel, often including sales and technical support.

Leveraging a third-party network and reseller type of agreements are generally simpler to achieve in most markets and involve extremely limited amounts of capital and resource investment by the entrant. In fact, in all but the rarest cases, there will be no in-country financial investment required on the part of the entrant and no in-country personnel. A simple technical interconnect with the in-country network (and a commercial agreement covering the same) is generally all that is required.

8.2.1 Common attributes of structures where entrant is the service provider

In terms of the form of entry where the entrant is the provider of telecommunications services, there will be several common attributes. The most notable of these are:

- *The need to create a legal entity.* New entrants, as discussed previously, are required in most countries to establish an in-country legal entity in order to hold a regulatory authorization and meet other government requirements, including those associated with tax and human resources requirements. The most notable exception to this general rule is the European Union, where one must only have a legal entity in a single member state to obtain a regulatory authorization and operate in other member states. Another time when a legal entity may not be mandatory is where a telecommunications service is unregulated and no regulatory authorization is required; such is often the case with e-mail, facsimile, voice mail, and other similar services.

  Even in the rare cases where a legal entity or local presence is not mandated, the market entrant may have other reasons to establish such a presence. First, tax considerations often drive structure. Failure to use a local entity may result in serious implications, such as the opening of a foreign corporation's books to the local tax authorities. Second, if personnel are to be hired and employed locally, it is generally either required or recommended, because of liability, tax, and other issues, that a local entity is used (another option is to hire the employees as contractors). Third, creating an in-country legal entity may insulate the entrant's parent from other liabilities, such as liability for certain faults. Accordingly, it is generally a good practice for a new entrant to establish an in-country legal entity once it makes an affirmative decision to enter the market.

- *Ability to obtain a regulatory authorization for the provision of services.* In order to deploy a network and provide telecommunications services as a service provider, most countries require some form of regulatory authorization. The form of the regulatory authorization can range from a basic service authorization that has numerous requirements and commitments, by which the service provider must abide, to a mere registration requirement in some countries for value-added services. In only limited cases can an entity provide telecommunications services without an authorization. An example of this is the provision of e-mail service in Brazil, which is not subject to a regulatory authorization requirement.

In order to obtain a regulatory authorization, an entrant may have to meet certain minimum qualification requirements. These requirements include financial and technical qualifications and may also include a certain level of skill as an operator, a demonstration of the availability of capital, and the like. In addition, the applicant may need to demonstrate that it has a plan for rollout of its services. The time frame for obtaining such an authorization may vary from around 30 days for a registration to several years in some of the more closed markets.

- *A location to build its network and access to essential facilities.* For service providers that are planning to place a network in-country, it is essential to have a location to house equipment. In addition, access to essential facilities, such as cable, and often other real property, such as roof tops for antennas or rights of way to lay cable, are important. Obtaining access to these resources generally requires the negotiation of commercial agreements with third parties or agreements with local governments.

  Often, the new entrant may meet with some resistance from these third parties. For example, it is not uncommon for the owner of a roof that is a critical antenna site to ask for large amounts of money and what may seem to be unreasonable commercial terms because of the importance of that site. Some countries, in an effort to eliminate such abuses, have adopted regulations that govern the more contentious portions of such negotiations. Such regulations must be evaluated on a case-by-case basis.

  Further, entrants may have to negotiate complex commercial agreements with the government to obtain access to real property. Complex applications may have to be prepared and participation in government hearings may also be necessary. These may be expensive and time consuming and should be planned on in advance by the entrant.

- *The need to utilize at least a part of the network of another service provider.* The very nature of telecommunications services requires access across multiple networks to provide the service. For example, for even a simple data service communication from Paris to London to be transmitted, the communication will involve the use of an international cable, local access capacity on both ends, and will need to hook through an in-country node or switch. In many cases, these functions can be contracted out to third parties. Each of these pieces of the communication may be carried by separate service providers. Even in the case of entry as an incumbent service provider, at some point, when competition takes hold, the service provider may need to utilize another service provider's network. For example, in Argentina, during its period of a duopoly for domestic service, both Telecom and Telefonica utilized each other's networks to carry the communication to the other service provider's customers.

In many cases, these negotiations are quite lengthy and contentious and often result in complex commercial agreements. In most countries, the negotiations with at least the dominant service provider for access to essential facilities will be governed by a set regulatory procedure. Unfortunately, despite such rules, the dominant service provider often is able to forestall competition by stalling negotiations or arguing other points that fall outside the gambit of the regulations.

- *The need for in-country personnel.* It is extremely important for both practical and legal reasons to have in-country personnel to run the telecommunications operation, handle commercial issues that may arise, sell services, interface with the government and customers, and be the in-country point of contact for the particular country. Such personnel can handle the day-to-day operational issues and be the face of the company to the government and customers. In addition, in many countries in order to obtain certain types of regulatory authorizations, the applicant must demonstrate that it will have qualified personnel in-country.

8.2.2 Common attributes of structure where third party is the service provider

In regard to the entry mechanism where the network of another service provider is being utilized by the new entrant, the following common attributes will generally exist:

- *A commercial agreement governing the relationship of the supplier and the entrant.* The nature of the commercial agreement that will govern the relationship between the entrant and the service provider is dependent on the service that will be provided, the technical operating characteristics of that service, the regulatory regime pursuant to which the supplier operates, the legal requirements of the country for that service, and the financial arrangements to which the parties have agreed. Accordingly, in each case, an individually tailored agreement will need to be concluded that addresses the broad aspects of the commercial relationship between the parties.

- *The supplier will need to hold the appropriate regulatory and other government authorizations to provide the relevant services.* Since the entrant has either decided against obtaining a regulatory authorization or believes that it may not be possible to obtain one, the entrant will generally rely on the supplier's regulatory authorization. Accordingly, the supplier and the entrant will need to have a meeting of the minds on the terms and conditions of use of the authorization(s).
- *The entrant may be limited in the manner in which it advertises services.* Quite often in the case of utilizing a third-party network and operating pursuant to its regulatory authorization, the entrant may not be allowed to hold itself out as the service provider. For example, in regard to the provision of reselling an underlying service provider's services, either the supplier or the regulatory body governing the supplier may require the entrant to advertise the underlying service as being provided by the supplier. This may impact the ability of the entrant to directly promote its services to its customers and even to enter into direct commercial agreements with customers for these services. For example, the entrant may need to state in its advertisement that the underlying telecommunications service is provided by the supplier and not the entrant.
- *The entrant may be limited in its sales territory.* In some countries, resale of another party's service is not permitted or requires a regulatory authorization, which may not be available to the entrant. Accordingly, sometimes a reseller may be limited in where it can offer the relevant service. For example, many parties are only authorized by third parties to sell and provide such services to customers with headquarters not located in the country or may only be able to provide the foreign half of the service.

8.3 Considerations specific to each of the five basic forms of entry

The following text is a more specific discussion of considerations that the entrant must take into account when determining what form of entry into a specific market is most appropriate to achieve its goals. As will be seen, these considerations may provide a greater incentive for an entrant to pursue a specific entry mechanism.

8.3.1 Full operating company

Entry into a market through a full operating company is generally fairly complex and involves the investment of substantial capital, time, and resources. Often, from the time of deciding to enter in this manner and being able to begin full operations, a lapse of several months to several years will occur, because of the need to obtain regulatory and other operating authorizations, access to facilities, and creating a network. This type of entry may occur either by entering through a new operation or the acquisition of an existing operator in the market. However, if an acquisition of an existing service provider is pursued, the time lag in terms of actually beginning operations will not exist, since the entrant will be acquiring a going company.

Another important consideration is the amount of government scrutiny that entry through the particular mechanism will create. Of course, in crafting this analysis, it is imperative for the parties that will be involved to be specifically addressed. For example, in many cases, additional levels of government scrutiny will exist for the acquisition of an existing operator by a large global service provider than by a small operator or a nontelecommunications service provider. Many countries mandate an antitrust review. This may also include extraterritorial application of trust laws by other jurisdictions, such as the European Union, where the entrant or service provider being acquired operates and has a minimum financial turnover in the subject market.

If an incumbent is being acquired, an additional important consideration is whether this occurs at the privatization stage or if one of the incumbents has already been through the privatization process. If the investment occurs at the privatization process stage, the entrant will likely have to compete in a competitive bidding situation with other potential service providers. However, through such a process, the entrant may be able to gain certain concessions from the government regarding its operations.

If, however, the incumbent service provider is being invested in a postprivatization process, the entrant will be more likely to face greater regulatory and antitrust scrutiny. Similar considerations may also exist if an established nondominant operator is the chosen entry mechanism. However, because of a lack of market dominance, generally the antitrust review and regulatory scrutiny will not be an entry hurdle.

No matter what method of entry is pursued for a full-service operating company, there are certain basic considerations that must be considered:

- The ability to obtain a full regulatory authorization for basic network services;
- The ability to build facilities and obtain interconnection from other service providers;
- The ability to have sufficient funds in which to make investments in infrastructure and network builds;
- The commitment to meet build-out and service obligations that may be imposed by the government;
- Establishing an effective billing collection mechanism;
- Expertise in the telecommunications arena.

If the entrant cannot meet any of these considerations, then the entrant may be well advised to rethink its proposed form of entry.

A good example of a service provider that has successfully deployed full operating companies is WorldCom, Inc., in Western Europe. As the European Union markets for telecommunications opened to competition, many companies saw a good opportunity to move into these markets. One of the most aggressive of these companies was MFS (ultimately to be acquired by WorldCom). MFS believed that by obtaining 100%-owned competitive regulatory authorizations in these markets it would be in a position to invest the funds and build out its own infrastructure (as well as lease the infrastructure of other service providers) to serve businesses located in key cities. To date, WorldCom has been successful and has a full facilities-based operation in almost every major EU member state, including the United Kingdom, France, Italy, Germany, and Spain [1].

8.3.2 Joint venture

In many cases, a new entrant may consider entering a new market with other service providers or partners (such as banks or energy companies). The decision to do this may be based on regulatory considerations, such as foreign ownership restrictions, or practical considerations, such as the need for increased capital or operating expertise. The joint venture is generally utilized in situations where the entrant wants to establish either a full operating company or a more limited network operating company.

A joint venture is often the most difficult of the entities to form. Complex commercial agreements, which govern operational, financial, and other aspects of the business, must be entered into between the partners. Such agreements typically address the following points:

- An allocation of responsibility for each operational area;
- How the management of the company will be established;
- How profits and losses will be allocated;
- Future plans for the company;
- A decision-making plan in case the partners reach an impasse on an important issue;
- A dissolution plan in case the joint venture fails.

If a joint venture is the preferred method of entry, there may be multiple entities formed—one to hold each investor's interest locally and then the local joint venture entity. This separation is often required to meet the legal and tax requirements of the jurisdiction where the entity will be operating.

There are many examples of both successful and unsuccessful joint ventures in the telecommunications arena. Many successful joint ventures have been formed by investing in new service providers, such as the partnership between Cable & Wireless, BellSouth, and others in Optus, Australia's second largest basic service provider.

Important aspects for a successful joint venture include:

- Compatible corporate cultures;
- Shared goals and vision of the market;

- A firmly mapped out business plan, which has flexibility built into it to adapt to the changing telecommunications marketplace;
- A simple, but effective management structure, where both parties believe that their interests are well represented;
- The combination of entities that have distinct operational responsibilities and strengths.

Unsuccessful joint ventures have included the previous incarnation of Global One, which, prior to being fully acquired by France Telecom, was a joint venture that consisted of France Telecom, Sprint, and Deutsche Telekom. This venture had as its mandate entering new markets primarily for data services, such as Frame Relay and ATM. In order for Global One to be formed, the proposed venture had to be reviewed and approved by both European and United States antitrust authorities and regulators, with certain operating conditions agreed to by the parties.

At first this venture appeared successful. In many of the markets where the Global One venture first operated its target market, business customers were clamoring for these advanced data services. Unfortunately, many of these markets were high cost or, where costs were lower, were very competitive, so that margins have not traditionally been high for Global One for these services. In addition, the differing cultures of the venture partners, or how and where the venture operated, began to impact the success of the venture.

Over time, growing dissension between the parties occurred and ultimately both Sprint and Deutsche Telekom were bought out by France Telecom. France Telecom also acquired Equant, an extremely successful global data service provider, and it now controls an increasingly large portion of the market for these services [2].

8.3.3 Limited network entry strategy

Another possible form of entry is a limited network entry strategy. Under this approach, the new entrant may only deploy a limited network, such as would be used for enhanced or value-added services. In general, as discussed, this type of approach requires a much more limited capital expenditure, because the equipment that is involved is much more limited in scope and there is less chance of a requirement to commit to the government to proceed with a large network build. In addition, regulatory scrutiny is generally limited, because of the smaller size of the operation and the less-pervasive services that are involved. However, when a dominant service provider is involved, there may be additional safeguards imposed on the operation of that service provider by the regulator to protect against anticompetitive conduct.

A good example of such a limited network strategy is the introduction of data services into a country. Quite often, the only expenses that will accompany such an entry mechanism will include:

- The cost of the network equipment;
- The cost of local and international capacity to provide service that is leased from third parties;
- The collocation fees to house the equipment and fees for any support;
- The fees of any in-country personnel to support the provision of service (for limited entry though, this should only comprise a few people);
- The cost of any fees associated with obtaining regulatory authorizations;
- Any taxes and any other costs of doing business.

Further, this type of operation is generally quick and easy to establish. First, the regulatory authorization is generally significantly easier and quicker to obtain than a basic service authorization. In addition, the network build tends to be rather simple, because of the limited scope of equipment. Often, the longest lead time and greatest expense in the effort is obtaining the bandwidth from third-party service providers.

Nonetheless, this approach is not hassle-free. Some of the problems with such an approach occur because of the limited presence of the service provider in the country. For example, from a regulatory perspective, because the entrant hires few or no personnel, has limited investment, and is a minor force in the market, it may be harder to garner the ear of the government due to anticompetitive conduct concerns on the part of the incumbent service provider. In addition, since limited entry means leasing space and technical space from third parties, the entrant is often reliant on these parties for the success of its operation. Often, these partners are direct competitors and may be fully prepared to meet their commercial obligations. Accordingly, it may be difficult for the entrant to make significant inroads into the market, at least on an expedited basis.

An example of this type of service provider is Equant, which was recently acquired by France Telecom. Equant has created a global data and IP network through the acquisition of entities with nodes in various countries and arranging partnerships with third parties to host their equipment. To date, Equant has one of the largest data and IP networks in the world [3].

8.3.4 Leveraging a third-party network

Another entry option, which is often quite attractive to companies when they cannot yet obtain a regulatory authorization or where the entrant is not yet comfortable deploying its own network in a market either for financial or other reasons, is to leverage a third-party network. Often, this type of arrangement also involves the entrant providing the third party with additional telecommunications equipment over which it will offer the services that the entrant wishes to provide in the country and abiding to specific service levels and operating requirements that the entrant is promising its customers it will provide. This form of entry is generally used for more limited services, such as the provision of data or other value-added services.

There are several major problems with this approach. First, expansion of sales into the relevant market is generally prohibited because of regulatory restrictions. Accordingly, sales must mostly occur in other countries and are mostly sold to global customers.

Second, because it is a third-party network, the quality of the network is dependent on the third party being able to meet the quality and consistency required by the entrant. In many cases, this is not possible, because the network standards vary so wildly internationally. Accordingly, service levels that are able to be provided may not be consistent with higher-end customer requirements.

Third, as the market opens, it may be difficult for the entrant to migrate its customers smoothly to its own deployed network because of existing contractual commitments. Further, such agreements often involve complex commercial arrangements. This often means that operational flexibility is lost through the bureaucracy of operating with a partnership or other commercial arrangement.

Many of the advanced data service companies, such as Global One, Concert, and WorldCom, operate through such arrangements in many countries [4]. These companies will partner through a commercial agreement with an in-country service provider, provide the high-technology equipment and expertise for the services they are looking to offer in the country, and allow the in-country service provider to be the operator of the service. These agreements may be based on a straight fee for service or may include some sort of revenue-sharing arrangement. In addition, the foreign service provider may obtain, in exchange for its services, the ability to access the global service provider's network around the world.

8.3.5 Reseller agreement

A close relative to leveraging a third-party network, but one that is more common, is the reseller agreement. This is differentiated from the form of entry of leveraging a third-party network, because it does not involve the same sort of involvement by the entrant. In this form of entry, generally a standard offering is created that is offered to numerous entrants on the same terms and conditions.

Resale provides an important source of competition in markets in which telecommunications infrastructure costs are high. Further, even new entrants will often need to interconnect their facilities with the incumbent's network to ensure wide service.

There are several types of reseller agreements. The three major types are:

- *Network-to-network interconnect agreements (NNI):* An NNI is an agreement between two service providers either domestically or internationally that connects two independent data networks for the purpose of providing connectivity between two or more user ports residing on those separate networks.
- *Straight resale agreements:* These agreements allow a service provider that wants to operate on a nonfacilities basis to obtain services from a third-party service provider and resell them to their own customers.
- *Correspondent agreements:* A correspondent agreement is an international agreement whereby each party carries international traffic on a half-circuit basis and settles for the differences in traffic based on agreed-upon accounting rates.

In regard to pricing, such services are economical to resell, because they are priced to the reseller based on wholesale pricing, because of the bulk services they are purchasing. However, many of the issues associated with leveraging the third-party network exist here as well. In addition, as the entrant steps farther away from having any control over the telecommunications operation, the quality of service to the entrant's customers often is degraded. Accordingly, such an approach may be less than optimal over the long term.

8.4 Continuation of case study: form of entry

MarathonTel management has decided to proceed with entering into the newly opened Eastern European market examined in Chapter 7. Despite concerns over the ability to obtain capacity on a timely and cost-based basis, MarathonTel believes customer demand is large and the market, over time, will develop an effective regulatory regime because of its relationship with the European Union.

The next decision that MarathonTel's management must make is what form of entry it should pursue. Because of the need for MarathonTel's management to retain control over its operations, a joint venture is not the optimal form of entry. In addition, because of the liberalization of the market, MarathonTel believes it is best to operate its own equipment in-country and enter into agreements directly to obtain international and local capacity for its services.

Countering the decision to have a full-blown operation, however, is MarathonTel's wish to limit capital investmenı and resources necessary for operation. In addition, because MarathonTel will only begin providing limited data services, a full facilities-based operation would not make sense. Accordingly, MarathonTel has decided it should pursue a more limited form of market entry, but one that utilizes its own facilities and ensures customers certain service levels.

In order to achieve this, MarathonTel has decided to:

- Provide and establish in-country equipment for data services;
- Rent collocation space for its equipment and obtain support for maintenance from the same provider;
- Establish an in-country shell entity and obtain a regulatory authorization for data services;

- Contract directly with international service providers for the international capacity for its services;
- Contract with a consultant to order local circuits from the in-country equipment to the local customer;
- Market services to global customers and then expand its reach to in-country customers, as the market matures.

References

[1] http://www.worldcom.com.

[2] http://www.francetelecom.com/vanglais/whats_in_the_news/f_news.html.

[3] http://www.equant.com/content/xml/nc_prehome_about.xml.

[4] http://www.concert.com/initial.asp.

Selected bibliography

Frieden, Rob, *Managing Internet-Driven Change in International Telecommunications*, Norwood, MA: Artech House, 2001.

9

Structuring a market to attract investors

9.1 Overview

This chapter is intended to provide national policy makers with an overview of some of the major policy options that are available for regulating telecommunications in markets that are seeking to create an effective competitive telecommunications sector. This chapter does not seek to mandate a specific solution but lays out viable options that are available today to regulators in creating, implementing, and enforcing such a structure.

In general, in order for policy makers to choose wise approaches for structuring their individual telecommunications markets, they must identify:

1. The goals for the telecommunications market from an economic and development standpoint;
2. The goals for the public in regard to telecommunications from a policy point of view;
3. The priorities that need to be set to achieve these goals and which goals have the highest priority.

Only once these goals and priorities are identified can appropriate regulatory structures and procedures be created in order to help achieve the best possible regime that will lead to the creation of a competitive market environment for that country.

In all cases, however, the government must also keep in mind the two main reasons that businesses enter new markets—to serve their customers and make a profit. Accordingly, corporations rarely have the same public policy incentives that governments do in creating competitive telecommunications legal and regulatory regimes. This fact cannot be ignored in developing any course of action for creating a competitive telecommunications market and will often drive the creation of safeguards and incentives that are necessary in order to achieve government goals.

Of course, the goals for the telecommunications market that the government identifies as key will not be stagnant, nor should they be. Accordingly, any regulatory and legal structure that is created to address the telecommunications market must be flexible and adaptable to the changing priorities and goals of the particular market. For example, in the United Kingdom, the initial goals of Margaret Thatcher's liberalization program in the 1980s included improving the United Kingdom's economy. The United Kingdom has expanded this and its other goals by encouraging new technologies, among other things [1]. One such example of this is the recent early auction of spectrum for third-generation mobile services. Throughout this time frame and the growing competitive nature of the telecommunications industry, only a small portion of the laws and regulations governing the United Kingdom's telecommunications market have needed to be amended or replaced to meet these evolving goals. This is because the United Kingdom created a flexible regulatory regime governing its telecommunications market. One of the major changes that has occurred, however, was a restructuring of the regulator to be more responsive to the changing nature of the telecommunications industry.

Generally, although new policies and rules may need to be enacted to execute competitive policy and other goals, many of the earlier policies and regulations remain in place and should continue to be effective at ach... the current goals of the government. Accordingly, it is ... the government ensure that its policies and rules can ...nging competitive environment over time. Another good ... flexible regime is in the United States. The underlying ...ications law in the United States dates from 1934 [2].

Although it has been amended numerous times to take into account new technologies and the changing market, the underlying foundations of the law and many key provisions, including its purpose, have remained intact and are as relevant today as in 1934.

The principal purpose of the United States Communications Act of 1934, as amended, is:

> To make available, so far as possible, to all the people of the United States, without discrimination on the basis of race, color, religion, national origin, or sex, a rapid, efficient, nationwide and worldwide wire and radio communications service with adequate facilities at reasonable charges, for the purpose of the national defense, for the purpose of promoting safety of life and property through the use of wire and radio communications, and for the purpose of securing a more effective execution of this policy.

This goal has remained the guiding principle in telecommunications policy in the United States over the past 7 decades. However, that has not meant that the U.S. government has not endeavored to meet other goals. These other goals have included:

- Infrastructure development;
- Education;
- Fostering a competitive telecommunications market;
- Universal service;
- Public safety and emergency preparedness;
- The free flow of information and ideas;
- Economic development;
- Encouraging the development of new technologies.

These goals have been achieved with only a few major rewrites of existing laws. The major reasons for these rewrites have been to allow increased competition into the U.S. telecommunications market. The most notable example of this was the Telecommunications Act of 1996, which expressly authorized competitive entry into the local service market and provided safeguards before certain incumbent local service providers could enter the long-distance market [3]. Having such flexibility in laws and policies allows the regime to grow freely without having constant changes being made that may be difficult to obtain approval for and may take extensive time to progress through government bureaucracy.

In addition, flexibility allows the regulatory regime to adjust to technological advances. For example, regimes such as that in the United States, have created regulatory regimes that are technology neutral. In addition, this concept of technology neutrality is being embraced by more and more countries and was a guiding principle during the negotiations of the WTO Basic Agreement. Accordingly, by ensuring technology neutrality, new entrants can retain the discretion to deploy the most appropriate technological solution for the relevant market and service.

Further, it is wrong for a government to believe that simply adopting core principles and approaches will lead to the creation of an effective competitive telecommunications market. It is important that the concept of an effective competitive telecommunications market be fully socialized to the community, both consumers and businesses, and receive some form of acceptance. Such acceptance of the legitimacy of the regime is critical to the success of the process.

In addition, as discussed, the government must be willing to ensure that anticompetitive behavior by dominant service providers or service providers that control essential facilities is curtailed. Failure to take action to remedy anticompetitive behavior that occurs despite the existence of core principles will lead to failure in the telecommunications market. A good example of this is the recent events in South Africa, where the South African government has failed to take action to reel in the anticompetitive behavior of Telkom in denying competitors access to its facilities, as required by South African law and regulation. This failure has negatively impacted consumers, since they have been denied the benefits of the limited competition that was authorized in the market. Further, the overall economy has suffered, since these service providers have not been permitted to grow their businesses, hence limiting economic investment and job creation. This has stymied a key effort of the South African government in economic development for the country.

Although the goals of countries often differ, finding common policy choices that many other countries also have followed have helped them achieve their distinct goals. Therefore, it is helpful if the government is able to study and discuss policy choices with regulators from around the world. Many regulators and other government agencies in markets where effective competition has been or is being successfully introduced have programs that regulators of foreign countries can tap into as a resource. Further, many of the international and regional organizations discussed earlier, such as the ITU, have programs for regulators that are developing and implementing competitive regulatory regimes. These regimes can be used as models for individual regulators to adapt to their own particular government, political, and economic structures.

9.2 Goals for the market

An important issue for the government to flesh out early on in the creation of its competitive telecommunications regime, and to reevaluate at regular intervals, is what the goal for the telecommunications market is in terms of economics and development. In order to identify these goals, an important threshold question is: What role does the government want telecommunications to play in its overall economic plan? The answer to this issue will directly impact how the telecommunications development plan is structured and implemented in that country.

If telecommunications were seen as a core industry for the country and an important magnet to attract and retain industry, then the government will have different goals and priorities than if telecommunications were seen as a less-strategic sector for the government. For example, if telecommunications is a driving force of the economy, then it is likely that lower rates for businesses and more city-based and higher-technology development will be encouraged than if the government sees the telecommunications industry in more of a role of a public utility. In the latter case, the government may believe that it is only necessary to encourage the development of the telecommunications sector into basic services but with a wider reach and on a more reasonable price basis throughout the country. Accordingly, the latter scenario may lead to stronger universal service policies, while the former may see incentives being provided to service providers that enter the market to provide higher-end services.

No matter what market goals are in place, there are certain key principles that must be met in order for a government to achieve the most basic of market and economic goals. These include permitting private investment, allowing foreign ownership, and creating a regulatory regime that is transparent and open in order to build investor confidence. Adoption of these principles will help develop a framework that will allow the market goals of a government to be met by encouraging private sector investment.

Private investment in the telecommunications sector is a key principle that governments should embrace. As discussed, most governments have traditionally operated and owned basic telecommunications service providers. However, privatization and liberalization of telecommunications markets began to become attractive, because of the incentives that private investors can bring to the marketplace. Private investors provide many things, including capital, expertise in marketing and other areas, and technological advancement. This often leads to improved service at more reasonable prices.

However, private investment must not just be allowed; it should be encouraged. Private investors will not enter a market where they cannot expect to earn a reasonable profit and have some form of operational control. Accordingly, governments should encourage private investment by ensuring that operational flexibility is provided to these operators and a regulatory regime exists that builds confidence. Although some governments may be hesitant to allow private ownership because of the control by a private entity of a strategic resource, these concerns can be addressed through the institution of regulatory requirements, which would ensure that private investors act in a manner that is consistent with the interest of that country. Such requirements may include instituting a universal service mechanism or establishing priority for certain communications, including emergency communications.

It is also important to allow foreign investment in the country for the purposes of developing the telecommunications sector. Foreign investment is a major source of capital in order to develop the infrastructure to provide modern telecommunications. Further, foreign investment often brings with it the technological and operating expertise that emerging markets are often lacking. Of course, as with private investment, some governments have concerns about allowing complete foreign ownership of strategic resources, such as telecommunications networks. Once again, regulatory requirements or limits can be imposed. For instance, there can be limits on the amount of ownership in a facilities-based service provider a foreign entity is allowed to hold. However, in structuring such limits, it is important to remember that many foreign capital partners believe it is imperative to be able to have operating control of the entity. Therefore, it is generally recommended that governments that limit foreign investment ensure that the foreign investor can have either de facto or de jure operating control of the service provider.

In order to attract private investment (both domestic and foreign) the marketplace must be more than just opened. It is imperative that the regulatory regime is also transparent and rational. Specifically, the other cornerstones, as discussed previously, that provide investors with confidence must also be met. These include enacting transparent regulations, creating an open and fair licensing process, establishing fair repatriation policies, and allowing investors to recognize a fair return on their capital investment.

9.2.1 Public goals

At the onset of establishing a competitive telecommunications regime, it is imperative that the government also map out its policy goals for the public from this process. Goals that are often identified include reasonable rates for consumers, access to telecommunications facilities for all citizens who desire it, and the introduction of advanced technologies.

The following principles are some of the key ones to having a government develop a competitive telecommunications market:

- Encouraging private investment.
- Providing open access to networks and services for providers and users.
- Creating a flexible regulatory framework to keep pace with technological and market developments.
- Universal service: Each country must define universal service for itself, in relation to its own goals, policies, and economic situation. Not only must geographic reach be addressed but also the services that should be encompassed within the universal service definition. Funding of universal service, on a neutral basis, is a key area that must be addressed.
- Competition policy: Anticompetitive safeguards must be imposed on the incumbent service provider and in the market in general to ensure that essential facilities are open to competition on a fair basis.
- Pricing regulation: Ensuring that prices are cost based, reasonable, and nondiscriminatory among similarly situated users.
- Encouraging new technologies: Rules should be developed in a manner that encourages innovations in technologies and allows

service providers the operational flexibility to deploy the technologies that are best suited for the service they are providing.

- Interconnection: Most incumbent basic service providers have no incentive to negotiate interconnection agreements with new entrants. Accordingly, regulations must establish key terms for interconnection, time frames for reaching agreements, and a neutral arbiter to arbitrate disputes.
- Unbundling: Incumbent service providers should also be required to unbundle their network facilities and services so that other service providers can utilize only the portions of the network they need to provide a service.
- Management of scarce resources: Access to scarce resources, such as spectrum, should be carefully managed to ensure that the resource does not lay fallow and that it is used efficiently.

Other public policy goals that the government may also seek to address include increased access to news and media, education, and telemedicine. Accordingly, the government may need to structure incentives for programs that will promote its public policy goals in these areas. Such promotion is not uncommon, and in some cases countries may provide incentives or require service providers to meet certain government goals.

Another issue that should be addressed is whether the government wants to provide preferences to certain historically disadvantaged groups, such as women or minorities. For example, in the United States, communications policy has often sought to encourage the participation of small businesses. This policy is well grounded in economic principles and in the concepts of fairness and equity. Accordingly, the government may wish to put in provisions that assist these entities in entering the market. These provisions may include providing financial credits in competitive bidding situations or setting aside a limited number of regulatory authorizations for members of such groups.

9.3 Determining government priorities

It is of critical importance for the government, once it identifies its market and public goals, to prioritize these goals. From a practical perspective, it is unlikely that a government will be able to achieve all or even the majority of its goals. Accordingly, while striving to meet each goal, it is helpful for the government to develop an action plan with a focus on meeting 5 to 10 critical goals and then endeavoring to meet the other goals over time. The European Union has done a very good job of prioritizing the goals for the Common Market's telecommunications liberalization program through the institution of its directives to member state governments concerning telecommunications [4].

Regardless of the regulatory model that a government adopts, its regulations, laws, and policies should clarify the respective rights and obligations of incumbent and new service providers. New entrants need regulatory assurances that the incumbent service provider(s) will not be allowed to use its dominant market position to hinder competition. Similarly, transparent and public regulatory processes create stable commercial environments, which encourage investment.

Further, it is imperative that governments not establish burdensome regulations. Traditionally unregulated areas should not be seen as prime targets for regulation, unless a need arises. This type of regulatory restraint is at least partially responsible for the success of the Internet. Yet a structure devoid of regulation, as discussed previously in regard to New Zealand, is not optimum. Regulation is necessary to achieve many of the goals of competition, including preventing anticompetitive abuses in the market. If the regulations that are adopted are flexible, they will more likely be able to adapt to the changing competitive environment. In addition, over time, the government should review its rules to make sure that they are not outdated or overly burdensome and make appropriate changes on a timely basis.

References

[1] http://www.oftel.gov.uk.

[2] 47 U.S.C. § 151 et seq.

[3] 47 U.S.C. § 251 et seq.

[4] Satellite Communications Directive, 94/64/EC, OJ L 128/15 (October 13, 1994); ONP Framework Directive, 90/387/EEC, OJ L 192/1 (June 28, 1990); Licensing Directive, 97/13/EC. OJ L 117/15 (April 10, 1997).

Selected bibliography

Intven, H., *Telecommunications Regulation Handbook*, Washington, D.C.: InfoDev, 2000.

10

Conclusion: where are we headed?

As we have seen in this book, many countries have already created or are in the process of creating competitive telecommunications markets. However, many more countries have been hesitant in this endeavor and have continued to keep their markets closed or only allowed minimal inroads to competition. Further, many of the carriers that have attempted to create, at least in part, competitive markets, have seen a recent downturn in these markets as the health of the global economy began to decline dramatically in the early 2000s. The future of the global telecommunications market has been led into even further turmoil with the tragic events of September 11, 2001, in New York; Washington, D.C.; and Pennsylvania.

Accordingly, there are three critical questions that are raised in regard to the future of global telecommunications markets:

1. Is the model of a competitive telecommunications market the right one for most, if not all, countries from a public policy perspective, in light of the downturn in the economy?
2. Will the downturn of the economy force the retrenchment of the telecommunications sector into larger service providers, and, if so, what will be the impact?
3. What does the future hold for the telecommunications market?

These three questions are discussed next.

10.1 The model of a competitive telecommunications market

Throughout this book, the basic regulatory formula that espouses the creation of a competitive telecommunications market has been reiterated, along with the benefits that it brings. However, the question still remains whether, from a public policy perspective, this is the correct philosophy for most, if not all, countries.

A survey of the markets around the world where true competition exists in the telecommunications sector demonstrates that in most cases it is the right approach from a public policy perspective. In most markets, consumers have directly benefited by increased access to telecommunications services, new and innovative services being introduced, and reduced rates being available. We have explored many examples of such benefits being received around the world. These benefits began to be recognized even during the recession-filled years of the early to mid-1980s.

Similarly, a competitive telecommunications market can help ensure that general government aims that most countries hold dear, including improving the welfare of their people and increasing the economic welfare of the country, are met through improving the ability of nationals to communicate. However, it would be dishonest to suggest that the institution of immediate competition in every country is always the correct approach. In some countries, it is appropriate to allow a phased approach to competition in order to encourage that certain individual market goals, such as network expansion and modernization, are achieved. In other markets, allowing full competition in all markets from day one may not provide the economic incentives to service providers to grow their networks. However, such a phased approach requires extreme diligence on the part of the regulator to ensure that a level playing field is created as greater competition is introduced into the market.

10.2 Impact of retrenchment

The early 2000s have seen a dramatic shift in the telecommunications market. Today more and more telecommunications companies are folding and consolidating after the gold rush years of telecommunications in the 1990s. It is likely that we will see a retrenchment in the telecommunications market because of the cyclical nature of telecommunications and the resource intensiveness of the industry.

As we discussed previously, during the early history of the telecommunications market, there were a multitude of service providers. However, as the general population of each country became more dependent on telecommunications services, and interconnection between networks was required, single service providers became the norm.

Over time, as even more innovative services became available, telecommunications had an even larger impact on society, and competition in the telecommunications market began to emerge. The 1990s began to see many telecommunications markets open their doors to at least some form of competition. At first, many of these new entrants were established service providers in other countries, such as Telefonica, France Telecommunications, and AT&T. As these entrants were able to develop market share and improve their financial earnings, other, less-established, entrants began to enter competitive markets. Often, these new service providers were ill equipped to handle the complexities of the telecommunications market, including fighting the large battles that often accompany being a new entrant in a market, such as obtaining interconnection with established incumbent service providers.

Further, many of these new service providers were casting themselves as Internet service providers or entities that supported Internet service providers. Unfortunately, when the gold mine of the Internet began to decline and many Internet service providers began to go out of business, these new entrants were also impacted.

New service providers have also been impacted. For example, many new technologies, such as mobile satellite services and point-to-multipoint microwave telecommunications services, have not resulted in the success that providers such as Iridium and Teligent had counted on. Accordingly, many networks dependent on these technologies have also been driven out of business.

With this downturn, many smaller or less well-funded service providers have gone out of business and/or sold their networks and customers to third parties. Often, these third parties have been the larger, more established service providers. Accordingly, the telecommunications marketplace is becoming a marketplace where the majority of market share is held by larger telecommunications service providers.

Despite this trend, in virtually every major telecommunications market where liberalization has occurred, there are still several major players emerging in the market and quite a few alternative service providers (such as for value-added, mobile, or other services) operating in the market. In less-liberalized markets, there may be few players overall. However, it is clear that even with the decrease in the number of service providers worldwide, as markets create improved competitive conditions, there will be an influx of several new entrants. These entrants will likely be better funded than many of the previous entrants and more committed to working to mature with the marketplace.

In addition, this will directly impact the need for the development of level playing fields in liberalizing markets. As service providers consolidate and as funding for expansion becomes more difficult to obtain, markets will have to attract investors. In order to do this, they will have to create attractive regulatory, legal, and policy frameworks for service providers. Failure to do so may result in a service provider not choosing to enter that market or to leave the market in order to save the expenses of fighting continual barriers to being able to provide services on a competitive basis.

10.3 The future of the telecommunications industry

Where do we go from here? If we could accurately answer that question, we would all be hailed as the great soothsayers of the twentieth century. Yet, there are certain events that are very likely to take place over the next 5 to 10 years. These include:

- The privatization of many of the government-owned service providers in the developing countries. However, it is likely that this process will see a slowdown because of the current state of the economy and the hesitancy of investors.
- The increasing liberalization of closed or partially closed markets, albeit, on a slower scale than we have seen in the past.
- The increasing importance of effective regulation as more and more incumbent service providers seek to retain market share as their home markets are opened to competition.
- Improved technology with a focus on the delivery of data services to capitalize on the success of the Internet.

- An increasing reliance on mobile communications with a focus on combined voice and data capabilities in an effort to save expensive fixed-line build-outs and to capture high-end users.

What is clear is that each event will be highly dependent on the others occurring. In addition, it is not clear what impact new technology will have on telecommunications. At the end of the day, it is clear that technology is a key driver in the development and growth of telecommunications markets.

It is important to note that the tragic events in the United States on September 11, 2001, have changed the world and will have a profound impact on the telecommunications industry. These events have highlighted the increasing dependence of the United States and other countries on telecommunications. In the hours and days following the September 11, 2001, incidents, it has been estimated that the highest level of international telecommunications traffic ever was carried. In addition, the sale of mobile telephones increased exponentially.

Despite these events, it is unclear what the future holds as the world prepares for a protracted battle against terrorism. For example, it is not certain how such a battle will impact the willingness of governments to further open their markets to competition and foreign investment. In addition, it is not clear how newly enacted security rules will impact the ability of service providers to import equipment necessary for telecommunications services into some markets. What is clear, however, is that the telecommunications industry will continue to be a dynamic one with a clear need for domestic and international public policy direction in order for it to develop in a manner that benefits the world's population.

List of Web sites

The following is an alphabetical list of available Web sites of entities that are mentioned in this book. These Web sites were current as of this writing.

Americatel: http://www.americatel.com.sv

Asia Pacific Economic Council: http://www.apecsec.org.sg

Central Provident Fund: http://www.cpt.gov.sg

Citibank: http://www.citibank.com

COFETEL (Mexico): http://www.cft.gov.mx

Compania de Telefonos de Chile: http://www.ctc.cl

CONATEL (Venezuela): http://www.conatel.gov.ve

Deutsche Telekom: http://www.dtag.de

Directorate General of Posts and Telecommunications (Vietnam): http://www.phsmou.or.jp

Empressa Nacional de Telecomunicaciones (Chile): http://www.entelchile.net

European Telecommunications Office: http://www.eto.dk

European Union, DG IV: http://europa.eu.int/comm/competition.net

FCC (United States): http://www.fcc.gov

Global Crossing: http://www.global.crossing.com

GTE (now Verizon): http://www.verizon.com

HONDUTEL: http://hondutel.hn

Hong Kong Telephone Company: http://www.hkt.net

Hutchinson Telecommunications: http://www.itahk.org.hk/profile/hutchinson.html

L'Office Conglaise: http://www.trasa.org/memberstate/congo/profile.html

MATAV Hungary: http://www.matav.hu

Netia Telekom: http://www.netia.pl

New World Telephone Hong Kong: http://www.newworldtel.com

Office of the Telecommunications Authority (Hong Kong): http://www.ofta.gov.hk

OFTEL (United Kingdom): http://www.oftel.gov.uk

Optus: http://www.optus.net.au

Organization for Economic and Cooperative Development: http://www.oecd.org

OTE (Greece): http://www.ote.gr

PanAmSat: http://www.panamsat.com

Proctor & Gamble: http://www.pg.com

Saigon Post and Telecommunications Service: http://www.saigonpostel.com.vn

Saltel: http://www.saltel.net

SATRA and ICASA (South Africa): http://satra.gov.za

SIGET (El Salvador): http://www.siget.gov.sv

Sprint Corporation: http://www.sprint.com

StarHub Singapore: http://www.starhub.com.sg

Swisscom: http://www.swisscom.com

Telecam: http://www.telecam.net

Telecom Colombia: http://www.telecom.com.co

Telstra Australia htpp://www.telstra.com.au

Temasek Holdings: http://www.tamasekholdings.com.sg

Undersecretariat of Telecommunications: http://www.mtt.cl

United States Department of Justice: http://www.doj.gov

USTTI: http://www.ustti.ort

Vianet: http://www.vianet.com.sv

Vietnam Posts and Telecommunications: http://www.vnpt.com

World Trade Organization: http://www.wto.org

WorldCom, Inc.: http://www.worldcom.com

List of abbreviations and acronyms

1996 Act	U.S. Telecommunications Act of 1996
APT	Asia Pacific Telecommunications organization
ATM	Asynchronous Transfer Mode
BOC	Bell operating company
CANTV	Compania Anonima Nacional de Telefonos de Venezuela
CITEL	Inter-American Telecommunications Commission
CONATEL	Comison Nacional de Telecomunicaciones
CPE	Customer premises equipment
CTC	Compania de Telefonos de Chile
DOJ	U.S. Department of Justice
EFTA	European Free Trade Agreement
EMBRATEL	Empressa Brasilera de Telecomunicaciones S.A.
ENTEL	Empressa Nacional de Telecomunicaciones
EU	European Union

FCC	Federal Communications Commission
FRS	Frame Relay Service
HKTC	Hong Kong Telephone Company Limited
HKTI	Hong Kong Telephone International
ICASA	Independent Communications Authority of South Africa
IP	Internet Protocol
ITU	International Telecommunication Union
MCI	Initially Microwave Communications Incorporated, later MCI Communications Corporation. In 2000, after it was acquired by WorldCom, Inc., the name of the company was officially changed to WorldCom, Inc.
MFJ	Modified Final Judgment
MFN	Most Favored Nation (under WTO)
OCPT	L'Office Congolais des Postes et Telecommunications
OECD	The Organization for Economic Cooperation and Development
OFTA	Office of the Telecommunications Authority (Hong Kong)
PSTN	Public Switched Telephone Network
PTT	Postal, Telephone, and Telegraph entities
SATRA	South African Regulatory Authority
SUBTEL	Chilean Undersecretariat of Telecommunications
TDP	Telefonica del Peru
TELEBRAS	Telecomunicaciones Brasileras S.A.
UN	United Nations
USTTI	United States Telecommunications Training Institute
VAS	Value-added services
VNPT	Vietnam Posts and Telecommunications

WTO	World Trade Organization
WTPF	World Telecommunications Policy Forum (ITU)

Glossary*

accounting rate A unit of currency negotiated by international carriers as the amount of compensation to cover the cost of completing an international call. This amount includes all satellite or submarine cable transmission costs and domestic tail circuits used to link an international gateway with the call originator and recipient. Carriers typically divide the accounting rate in half when settling accounts.

Additional Plenipotentiary (APP) An extraordinary Plenipotentiary Conference of the International Telecommunication Union held in 1992 to consider recommendations on how to streamline operations. The APP significantly revised the Constitution and Convention of the ITU.

Administrative Procedure Act An Act of the United States Congress setting forth required procedures that administrative and regulatory agencies must use to provide for public participation and due process in the setting of rules and policies.

ADR Alternative Dispute Resolution. The use of conflict resolution alternatives to litigation and conventional administrative channels, including negotiations managed by a private facilitator or mediator.

---

* Reprinted, by permission, from Rob Frieden, *Managing Internet-Driven Change in International Telecommunications*, © 2001.

ad valorem tariff A tariff calculated as a percentage of the value of foreign goods clearing customs.

advance publication The process by which nations inform other nations of future satellite orbital arc requirements through the submission of information about the satellite and the desired orbital location. The Radio Regulations Board, formerly known as the International Frequency Registration Board, of the International Telecommunication Union provides the forum for dissemination of orbital arc requirements and avoidance of resolution of actual or potential interference between satellites.

algorithm The application of a computing principle that sets the foundation for processing or transmitting data.

analog Information transmitted by varying modulation over a radio carrier wave. Voice and music originate in analog form and are converted for transmission over digital facilities like fiber optic cables.

ANI Automatic Number Identification. A feature in advanced telecommunications routing that provides the calling party's telephone number as the call is set up. Telephone companies have packaged this feature as a way to screen calls in residences and as a marketing and billing tool in commercial applications.

ANSI The American National Standards Institute. A nonprofit organization addressing standard setting, primarily by certifying other expert bodies to formulate standards in a narrow area of expertise.

APEC Asia Pacific Economic Cooperation group. An association of nations in the Asia/Pacific region that addresses trade and development work, including telecommunications and electronic commerce issues.

arbitrage A brokering function where a business acquires bulk capacity and resells it to individual users who singularly could not qualify for the discounts accruing to large volume customers.

Arianespace The world's first commercial space transport company established in 1980 by 36 European aerospace and electronics companies, 13 major banks, and the French space agency. The company launches satellites from a facility in Kourou, French Guyana.

ASEAN The Association of Southeast Asian Nations. A forum of Southeast Asian nations for addressing issues of mutual economic, technical, and trade issues. Many of the ASEAN nations have leased satellite capacity on the Palapa satellite system operated by the Indonesian government.

Asia-Pacific The region encompassing the Pacific Rim nations and countries bordering the Indian Ocean. This region has experienced the greatest economic growth that in turn has stimulated explosive development of telecommunication infrastructures.

Assembly of Parties The forum where representatives of nations, which have associated with the INTELSAT Agreement and Inmarsat Convention, meet to address major policy issues involving the cooperatives.

asynchronous transfer mode An advanced form of data transport whereby users can dynamically select the amount of bandwidth and transmission speed required and change the assignment as conditions warrant.

Atlantic Ocean Region (AOR) One of three major geographical regions, identified by a prominent body of water, that outline the largest possible satellite footprint. A satellite hovering above the equator midway between North America and Europe will have a footprint illuminating all of the Atlantic Ocean and most of North, Central, and South America; Europe; and Africa.

autonomy The assertion of independence and sovereignty which, if recognized by other nations, confers the right to participate in the International Telecommunication Union, INTELSAT, and other forums where policy and investment decisions are made.

balanced loading A regulatory policy requiring international carriers to activate satellite and submarine cables on a prescribed ratio or even basis with an eye toward bolstering use of the more expensive or developing transmission technology.

band A term that refers to a specific range of frequency spectrum. For example, the C-band refers to frequencies used to uplink and download traffic to satellites: around 6 GHz uplink and 4 GHz downlink.

bandwidth The total range of frequencies necessary to accommodate a spectrum using signal without distortion. Typically, bandwidth requirements grow as the amount of information to be transmitted increases. For example, an AM radio signal contains 10 kHz while an FM channel contains 200 kHz and a broadcast television signal contains 6 MHz, or 6,000 kHz.

bandwidth on demand The ability of users to adjust up or down the amount of capacity leased from a carrier, primarily through new technologies that make it easier to aggregate bitstreams.

barriers to market entry Structural, regulatory, and trade limitations to competition by domestic or foreign enterprises. Such barriers include prohibitions or limitations on the amount of foreign investment in a telecommunications service provider, reserving service monopolies to the incumbent carrier, imposing duties and other financial impediments on the equipment or services, and subjecting foreign carriers and manufacturers to more burdensome certification, testing, and licensing burdens.

basic rate interface The least common denominator of capacity in Integrated Services Digital Networks comprised of two 64 Kbps bearer circuits and one 16 Kbps data channel.

basic services The switching, routing, and transmission of voice or data provided by facilities-based carriers traditionally subject to common carrier regulation. Enhanced service providers use basic services as "plain vanilla" building blocks over which additional services and features are added.

bearer circuit A transmission pathway that can be subdivided into channels of less bandwidth or data throughput through multiplexing. For example, a bearer circuit with a total capacity of 64 Kbps can be subdivided into four channels with the capacity to transmit one voice grade channel.

beggar-thy-neighbor policy Attempts by one country to reduce unemployment and to increase domestic output by raising tariffs and erecting nontariff barriers to reduce imports. This strategy has proven risky because it can provoke retaliation.

Bell Operating Company (BOC) The telephone service operating companies of the regional holding companies divested from AT&T. For example, the regional holding company Verizon operates several BOCs serving customers from New England to Virginia.

bilateral trade agreement A formal or informal agreement between two nations.

bit "Binary digit." The smallest unit of measurement in the transmission of data having only two values called 0 or 1.

bit error rate The extent to which a digital transmission network generates an error. The rate can be reduced through error detection and correction.

bit rate Also known as "throughput," bit rate is the speed at which bits are transmitted, usually expressed in terms of bits per second. For example, a conventional dial-up modem can handle data at the rate of up to 56,000 bps.

BOO Build, Own, Operate. A method for infrastructure development where a foreign enterprise agrees to build and operate a facility. This arrangement creates a service franchise and the incentive for the operator to upgrade and maintain facilities without fear of nationalization.

boomerang box A device that provides dial tone to callers in another country thereby providing them a virtual presence in a location with lower outbound international long distance telephone rates.

boresight The center point on Earth where a satellite's radiated transmissions are strongest.

BOT Build, Operate, Transfer. A method for infrastructure development where a foreign enterprise agrees to build a facility, operate it for a specified time period, and then transfer title to the national government or carrier who will then take over operations after having had time to develop operational and management expertise.

bottleneck A facility or portion of a route where traffic aggregates, often allowing the operator or service provider the opportunity to charge monopoly rates and to engage in anticompetitive practices.

boycott A refusal to deal commercially or otherwise with a person, firm, or country.

broadband A medium offering high capacity, high speed services achieved by using large wide bandwidth and digital transmission.

browser Software that provides a convenient and easy-to-use interface for accessing content on the Internet.

BSA Basic Serving Arrangement. A term used by the Federal Communications Commission in its Third Computer Inquiry to identify the components in generic connections between Enhanced Service Providers and facilities-based carriers providing basic services. BSAs consist of the access links between facilities as well as the transport, routing, and functions.

BSE Basic Serving Elements. A term used by the Federal Communications Commission in its Third Computer Inquiry to identify optional network features, available on an unbundled, à la carte basis such as ANI.

BT An acronym referring to British Telecom, the dominant local and international carrier in the United Kingdom.

BTO Build, Transfer, Operate. A method for infrastructure development where a foreign enterprise agrees to build a facility, transfer title to the national government or carrier, and operate the facility at a profit for a specified time period after which a domestic carrier may take over operations. This arrangement enables the government to maintain a greater degree of facilities control than if a private foreign entity held title.

bundled The process by which two or more, possibly segregated, features are offered jointly.

bypass The use of alternative facilities or services to those of incumbent carriers (e.g., the use of cable television systems or wireless technologies to access interexchange carrier facilities rather than the incumbent telephone company's wireline network).

byte An intermediate unit of data transmission capacity comprised of 8 bits.

C3I Command, Control, Communications, and Intelligence. Tactical requirements of the military.

cablehead The oceanfront location where a submarine cable makes landfall and where power, amplification, and multiplexing functions take place.

caching Temporary storage of Internet content (e.g., frequently viewed World Wide Web pages at servers closer to a particular user).

callback services Providing outbound international long distance calling capabilities to customers in another country through the use of a device that processes a request for dial tone. Callback service can be initiated by a conventional inbound international long-distance call that is terminated before triggering a toll charge—the process of importing outbound telephone calling capability into a location having high domestic and international calling rates.

call-reorigination The process of reversing the routing of an international telephone call for purposes of providing consumers cheaper outbound calling opportunities.

C-band The portion of the radio spectrum (Earth-to-space at 6 GHz, space-to-Earth at 4 GHz) allocated for satellites providing service between fixed points or broadcast services.

CCIR The International Radio Consultative Committee of the International Telecommunication Union, reformulated in 1993 with portions of its portfolio assigned to the newly formed Radiocommunications and Telecommunication Standards sectors.

CCITT The International Telegraph and Telephone Consultative Committee of the International Telecommunication Union, reformulated in 1993 with portions of its portfolio assigned to the newly formed Radiocommunications and Telecommunication Standards sectors.

CDMA Code Division Multiple Access. A process for deriving more throughput and accommodating increasing demand for service by subdividing spectrum into various code sequences.

CEI Comparatively Equivalent Interconnection. A precondition imposed by the FCC before the BOCs could provide basic and enhanced services without structural separation.

cellular radio A terrestrial radio service designed primarily for mobile applications using microwave transmitters whose low powered operations enable frequency reuse and integrated service throughout a region.

CEN European Committee for Standardization. A regional standard-setting body in all fields except electrical and electronic matters.

CENELEC European Committee for Electrotechnical Standardization. A regional standard-setting body in electrical and electronic matters.

central office A telephone company facility that provides centralized management of switching, routing, and line transport functions that may traverse other facilities closer to the end user.

CEPT The Conference of European Post and Telecommunication Administrations. An assembly of European postal and telecommunications regulatory authorities aiming to harmonize standards, policies, and initiatives for the ITU and other global forums. Diversifying constituencies and establishment of a separate standard setting body (the European Telecommunication Standards Institute) challenge the ability of CEPT to establish significant and uniform policies.

circuit switching Physically linking two or more points via dedicated circuits.

CIS The Commonwealth of Independent States. An affiliation of sovereign states that formerly comprised the Soviet Union.

CITEL Inter-American Telecommunications Conference. An assembly of nations that participate in the Organization of American States with a mission to address radio and telecommunications issues, including ones to be addressed by upcoming conferences of the ITU.

collection rate The end user charge for service, typically set out in a tariff.

colocation The physical interconnection of lines and equipment owned and operated by different carriers typically on the premises of the major incumbent carrier.

commercialization of space Eliminating government subsidies and management so that private ventures can compete fully and fairly in space-related ventures.

common carrier A legal and regulatory classification that requires a telecommunications facility or service provider to serve any user within a certificated geographical region, and to provide service in a nondiscriminatory manner typically through public tariffs.

Communications Act of 1934 The primary United States law establishing the Federal Communications Commission and the general scope of broadcast and common carrier regulation.

Communications Satellite Act of 1962 The United States law that created Comsat Corporation as the sole signatory to INTELSAT and established general satellite policy. Congress amended this Act to permit the acquisition of Comsat by Lockheed Martin Corp., the privatization of INTELSAT, and more robust international satellite service competition.

comparably efficient interconnection The requirement of the Federal Communications Commission that the Bell Operating Companies provide a plan demonstrating that they will not discriminate against unaffiliated enhanced services providers when the BOC decides to provide similar services. While the Second Computer Inquiry required the BOCs to provide enhanced services through a separate subsidiary, the Third Computer Inquiry eliminated structural separations provided the BOCs filed a CEI plan for each enhanced service.

comparative advantage A fundamental international trade concept that views nations or regions having the ability to produce certain goods or services as being more efficient than other nations. By emphasizing production where a comparative advantage exists, nations can efficiently use available natural and human resources, and trade for goods and services where the nation has a comparative disadvantage.

compatibility The ability of users and carriers to interconnect equipment, lines, and facilities while maintaining services with a reasonable degree of reliability and quality.

complementary products Products that add to the value and utility of a product (e.g., sugar with coffee and modems with personal computers).

compression The application of techniques for reducing the amount of frequency spectrum or channel capacity needed to derive a circuit (e.g., 4 to 1 compression provides the means for deriving four channels where previously only one channel was available).

Computer Inquiries A set of proceedings of the Federal Communications Commission beginning in the 1970s designed to erect a regulatory system that permits facilities-based carriers to enter enhanced services markets without cross-subsidization by users of the carrier's basic services.

Comsat Corporation The former exclusive U.S. signatory (investor) in INTELSAT and Inmarsat that operated as a "carrier's carrier" by leasing wholesale satellite capacity to other carriers for subsequent retailing to end users. Lockheed Martin Corporation acquired Comsat in 1999.

conscious parallelism The deliberate matching of prices and services by carriers to avoid more aggressive competition and price wars.

consent decree A remedy in an antitrust case where the proceeding concludes without a verdict in exchange for an agreement by the defendant to refrain from continuing to engage in certain activities or practices. AT&T agreed to consent decrees in 1956 and 1982 to settle antitrust suits.

consortium A still-predominant ownership structure where multiple carriers from several nations jointly participate in the ownership of a submarine cable or satellite venture.

consultation process Scheduled meetings of international carriers and government agencies to evaluate facilities demand and determine where and when to deploy additional capacity.

consultative process The requirement in Article XIV of the INTELSAT Agreement and Article VII of the Inmarsat Convention that parties to these agreements consult with the cooperative to ensure that another satellite system will not cause technical or economic harm.

CONUS Continental United States. The lower 48 states.

convergence The merger of technologies that previously served discrete markets leading to integrated, additional offerings (e.g., computer terminals will serve entertainment applications in addition to information processing, and television sets will serve information processing applications in addition to entertainment).

cooperative The formerly predominant method for spreading financial and operational risk for operating regional and global satellite ventures.

coordination The process by which nations, carriers, and other service providers meet to resolve potential conflicts, including the potential for radio interference. The INTELSAT Agreement requires nations that have agreed to become parties to the organization to demonstrate the absence of technical and economic harm to INTELSAT when authorizing separate satellite systems. The Radio Regulations of the International Telecommunication Union require nations to coordinate the use of frequencies and the satellite orbital arc.

COPUOS Committee on the Peaceful Uses of Outer Space. A Committee of the United Nations that addresses issues pertaining to space exploration, settlement, and orbiting objects in space, including satellites.

corporatization The conversion of a government-owned carrier into a more business-like enterprise, with or without a change in ownership.

CPE Customer Premises Equipment. Telecommunications equipment, including telephones and private branch exchanges, located on user premises.

CPNI Customer Proprietary Network Information. Information about a customer's basic service usage that can provide marketing leads to enhanced service providers. The Computer Inquiries established rules requiring the BOCs to withhold such information from enhanced service affiliates or competitors unless authorized to do so by the user.

cross-subsidization Using revenues accrued from one service to underwrite the provision of another service at less than fully compensatory rates.

cyberspace A word coined by William Gibson in his novel Neuromancer now referring to the broad, global resources available via computer networks.

DAB Digital Audio Broadcasting. The transmission of audio signals in a higher quality digital format rather than the conventional analog method.

DAMA Demand Assigned Multiple Access. A queuing method by which a larger number of users may share transmission capacity by allocating on an as-needed basis.

DBS Direct Broadcast Satellite. The use of medium to high powered signals for direct broadcasting to satellite terminals located on the premises of the recipient.

dedicated Reserved for the use of one or more specified users (e.g., a private line or satellite transponder may be dedicated for the use by a particular lease). Nondedicated capacity is provided on a virtual or on-demand basis.

de facto standard setting The creation of a dominant or single standard by the interplay of market forces rather than promulgation of a standard through the standard setting process.

de jure standard setting The formation of a standard by law or through the rule-making process of the appropriate regulatory agency.

deregulation A reduction in the breadth and scope of regulation with an eye toward allowing marketplace forces to replace government intervention.

destructive competition Short-term price competition at noncompensatory levels resulting in the market exit by some enterprises and the potential for survivors to raise rates above competitive levels to recoup prior losses.

digital The use of a binary form, comprised of on and off pulses, to represent the continuously varying signals of images and sounds. Information, entertainment, voice traffic, and other forms of communication can be encoded, stored, processed, and transmitted in a numeric, as opposed to analog, form.

digitization The use of computer readable bit streams for transmitting, switching, and routing information.

direct access The opportunity for end users and service providers to acquire satellite capacity directly from the operator (e.g., INTELSAT and EUTELSAT instead of having to deal with an intermediary).

direct broadcast satellite The use of a satellite to transmit programming, including audio and video, directly to end users equipped with receiving dishes and the necessary electronic components (also known as direct to home broadcasting).

disintermediation The ability of new information services, like the Internet, to eliminate channels of distribution for a product or service.

divestiture Severing part of a corporation and creating a separate business entity, on a voluntary basis, or as part of an antitrust settlement or verdict.

DNIC Data Network Identification Codes. ITU-recognized coding system for identifying and routing traffic to specific network operators.

dominant carrier An FCC classification of common carriers who, because of their market power, require closer regulatory scrutiny. Such carriers have less flexibility to change tariff terms and conditions. The classification applies to AT&T, foreign-owned carriers on a route-specific basis, Comsat in its INTELSAT and Inmarsat signatory functions, and carriers operating the only Earth station in an offshore point (Guam, for example).

DOV/DUV Data-Over-Voice/Data-Under-Voice. A technique for transmitting voice and data over a single channel through multiplexing or other signal splitting technology.

downlink The transmission of traffic from a satellite to receiving locations.

duopoly A monopoly shared by two enterprises.

E-1 The European standard for transmission capacity handling 2.048 Mbps.

Earth station Terrestrial equipment used to transmit and receive satellite telecommunications. Some Earth stations provide receive-only functions.

economies of scale A measurement of economic efficiency that identifies who produces a good or service at the lowest per unit cost and the optimal amount of production.

economies of scope A measurement for assessing how efficiently an enterprise will operate when serving adjacent markets (e.g., whether a telephone company can efficiently use its network to provide cable television and information services).

EEC European Economic Community. A trade organization and governance structure for European nations now known as the European Union.

EFTA European Free Trade Association. A trading association of European nations not part of the EEC, including the Nordic Countries, Austria, and Switzerland.

elasticity of demand The intensity of preference for a good or service based on user reaction to an increase or decrease in price.

elasticity of supply How consumers react to changes in the availability of a good or service.

electronic commerce The use of the Internet and other data communications media to promote faster, better, smarter, and more convenient commercial transactions.

electronic data interchange The use of telecommunications and information processing to conduct business transactions, often in an integrated network combining different media (e.g., voice, text, and data processing).

electronic funds transfer The use of telecommunications and information processing to achieve a transfer of money, typically from one bank to another.

e-mail Computer-mediated text communications.

end office The telephone company switching facility closest to a particular user.

end-on-end routing The segmentation of a route into two or more segments typically priced individually.

end-to-end routing A complete traffic routing arrangement typically priced at one composite rate.

enhanced services Enhancements to basic common carrier transmission services involving computer processing that acts on the code, content, protocol, or format of the information in such a way as to change the output and possibly to store it as well for subsequent retrieval and manipulation.

enterprise networks Diversified, customized, and complex international telecommunications and information networks primarily used by multinational enterprises to serve particular requirements. These networks often require the assistance of system integrators and outsourcers who plan, procure, and manage the network.

Erlang A unit for measuring telecommunications traffic and capacity needed to support an acceptable level of service during the busiest usage times.

ETSI The European Telecommunications Standards Institute. A regional standard-setting body organized by the EU.

European Union The Union of European nations formerly affiliated in a less integrated European Community.

EUTELSAT The European Telecommunication Satellite Organization. A regional satellite cooperative providing satellites for delivery of telecommunications and video programming.

facilities-based competition Marketplace entry by enterprises that have installed their own transmission, switching, and related facilities rather than resell the facilities of another carrier.

FCC Federal Communications Commission. The expert regulatory agency created by federal law to allocate, allot, and assign spectrum and to oversee broadcasting and common carrier telecommunications.

FDC Fully Distributed Costs. Costing that includes a contribution to fixed costs by all consumers.

feeder link A radio link to and from satellites for conveying information, including the tracking, telemetry, and network control needed to maintain a satellite in proper orbit.

fixed satellite service The use of satellites to provide service between users at fixed locations.

footprint The range of geographical coverage of a satellite transmission.

forbearance The FCC-articulated concept of refraining from regulating certain common carriers lacking market power thereby affording greater flexibility in pricing and providing service.

foreign correspondent A carrier that has entered into a foreign operating agreement with another carrier entitling it to originate and terminate international traffic.

frame relay A new data transmission technology that quickly switches and routes digital packets with a low bit error rate.

gateway A satellite Earth station or submarine cablehead where domestic facilities access international transmission facilities.

GATS General Agreement on Trade in Services. The process for extending GATT trading principles to services in addition to goods.

GATT General Agreement on Tariffs and Trade. A multilateral agreement by most nations to reduce barriers to trade. The provisions of the GATT were integrated into the governance documents of the World Trade Organization.

geostationary The location above the Earth where launched objects appear stationary relative to Earth. Satellites become geostationary at

22,300 miles in altitude. Placing satellites in geostationary orbit above the equator maximizes the scope of geographical coverage.

geosynchronous The orientation of a satellite's orbit in relation to the Earth's orbit.

gigahertz (GHz) A measure of radio frequency equivalent to 1 billion hertz (cycles per second).

GII Global Information Infrastructure. A concept extending globally national initiatives to promote widespread deployment and access to broadband digital technologies and services requiring high speed digital transmission (e.g., video programming, telemedicine, large data file transport, and Internet access).

globalization Expanding the perspective, strategic business planning marketing, and operational scope of an enterprise to target and serve new international opportunities (e.g., PTTs, facing lost market share domestically as a result of market entry, may seek market share in foreign markets).

GNP Gross National Product. An aggregate measure of income generated from all goods and services.

green paper A vehicle for publishing future policy objectives. The Commission of the European Community prepared a green paper in 1987 articulating telecommunications harmonization goals as part of the Single Europe Act.

GSM Global Systems for Mobile Communications. A standard for digital cellular radio systems that originated in Europe.

half-circuit The smallest unit of international transmission capacity matched by foreign correspondents to route traffic.

harmful interference Spectrum usage that endangers, degrades, or deteriorates the proper functioning of another registered spectrum use.

HDTV High Definition Television. The development of a video transmission and production standard calling for higher resolution and better sound quality.

Heads of Agreement A preliminary agreement by governments and business enterprises articulating general terms followed by a more comprehensive document.

HLC An ad-hoc High Level Committee created by the ITU Plenipotentiary to provide recommendations to make the Union more efficient and effective.

hosting Serving as a repository for content made available via the Internet.

hubbing The aggregation of traffic from throughout a region thereby achieving circuit loading efficiency and economies of scale. Hubbing in telecommunications can be analogized to airport hubbing where long haul routes are loaded from a number of short haul traffic originating in nearby localities.

ICANN Internet Corporation for Assigned Names and Numbers. An organization seeking to establish uniform policies to resolve conflicts relating to the structure and names used to identify Internet sites.

ICAO International Civil Aviation Organization. A specialized agency of the United Nations that establishes rules, standards, and policies affecting international aviation, including navigation and telecommunication issues.

ICO Intermediate Circular Orbit. The installation of satellites into nongeostationary orbits closer to Earth to support applications using small, low powered handheld terminals.

IEEE Institute of Electrical and Electronic Engineers. An international organization that participates in the telecommunications standard-setting process.

IFRB International Frequency Registration Board. The part of the ITU's Radiocommunications Sector, now known as the Radio Regulation Board, that registers spectrum and orbital arc uses and resolves potential interference problems.

IMTS International Message Telephone Service. The conventional dial-up international long distance telephone service.

in-band signaling The customary process of including switching and routing information as headers preceding the call, increasingly replaced by out-of-band signaling.

incentive regulation Alternatives to the conventional rate base regulation designed to reward innovation and efficiency by allowing carriers to capture financial gains rather than automatically flow them to ratepayers through refunds or lower rates.

inclined-orbit Satellite orbits that conserve station-keeping fuel to extend the operational life of a satellite by allowing it to deviate slightly from a geosynchronous orbit, thereby requiring Earth station tracking.

incumbent carriers Carriers that have installed an extensive network and heretofore have faced limited, if any, competition.

infrastructure An essential system to the public health and welfare (e.g., water, sewerage, electricity, currency, telecommunications, and roadways).

Inmarsat International Maritime Satellite Organization. A cooperative of over 70 nations organized to provide ubiquitous maritime services, including safety and distress functions. The parties of Inmarsat expanded Inmarsat's function to include aeronautical and land mobile services on an ancillary basis. Inmarsat expects to privatize, having already spun off part of its assets into a competitive venture.

intelligent network The architecture and plans for a future telecommunications infrastructure with advanced features made possible by expanded network switching, signal processing, and intelligence.

INTELSAT International Telecommunication Satellite Organization. A cooperative of over 120 nations organized to provide ubiquitous international satellite capacity. INTELSAT expects to privatize, having already spun off part of its assets into a competitive venture.

interactive The ability of users to request, manipulate, process, and change data through on-line commands. Interactivity converts one-way information sources into two-way media.

interconnect The physical connection of equipment and lines to secure a complete route and service arrangement. Users physically interconnect CPE to a telecommunications network through a plug physically attached to a jack.

interexchange telecommunications Long haul services that link local exchanges. In the United States, the Modification of Final Judgment precludes Bell Operating Companies from providing those interexchange services that cross local access and transport boundaries.

interface A shared boundary (e.g., between CPE and the telecommunications plant owned and operated by a telephone company).

international comity Recognition that each nation has sovereignty and that mutually beneficial results accrue when nations relinquish a degree of independent decision-making to reach a single, consensus policy.

Internet A multifaceted, international network of databases and users, organized initially for military, scientific, and academic users.

Internet service provider An enterprise that owns or leases facilities to provide end users with access to the Internet. ISPs also may provide or host content.

Intersputnik The former international satellite cooperative organized by the former Soviet Union.

INTUG International Telecommunications Users Group. A trade association of international telecommunications users.

IRAC Interdepartment Radio Advisory Committee. A committee, organized under the auspices of the Commerce Department and comprised of 20–25 representatives from various U.S. federal agencies, with a mission to coordinate on spectrum management issues affecting the U.S. government.

IRC International Record Carrier. International carriers that concentrate on providing textual services like telegraphs and telexes.

IRU Indefeasible Right of User. A method for conveying the rights, but not the title, to use international telecommunications capacity.

ISDN Integrated Services Digital Networks. A telecommunications standard envisioning interconnected digital networks capable of simultaneous delivery of voice and data services.

ISL Inter-Satellite Links. Extremely high frequency communication links between satellites that substitute for a round trip from a satellite down to an Earth station and back up to another satellite.

ISO International Standards Organization. A specialized agency of the United Nations with a broad standard-setting portfolio that includes data processing.

ISP International Settlements Policy. An FCC policy requiring all U.S. carriers to apply identical accounting and settlement rates for IMTS. The FCC considered the ISP essential to prevent foreign carriers from whipsawing U.S. carriers, but now permits deviation from the ISP by WTO member nations lacking government monopoly carriers.

ITU International Telecommunication Union. A specialized agency of the United Nations that formulates policy, regulations, and recommendations in telecommunications, including spectrum allocation, satellite orbital arc registration, telecommunications development, standard setting, and conflict avoidance/resolution.

ITU Council An elected body of the ITU that performs executive board functions including the scheduling of conferences.

IVAN International Value Added Network. An enterprise that enhances leased private lines with customized features (e.g., credit card verification and airline reservation systems).

IXC Interexchange Carrier. A provider of long haul telecommunicatiosn services.

Ka-band That relatively unused portion of the radio spectrum (30 GHz uplink, 20 GHz downlink) allocated for new satellite requirements.

KDD Kokusai Denshin Denwa. The dominant Japanese international carrier.

kilohertz A measurement of radio waves equal to 1,000 cycles per second.

Ku-band An increasingly used portion of the radio spectrum (14 GHz uplink, 11–12 GHz downlink) allocated for satellite requirements.

LATA Local Access and Transport Area. A geographical region sharing common cultural, social, and economic interests within which the divested Bell Operating Companies may provide local and interexchange services.

L-band That portion of the frequency band (1–3 GHz) where a variety of mobile services are provided by geostationary and LEO satellites.

LDC Lesser Developed Country. A classification of nations with economies and telecommunications infrastructure below that of industrialized nations. Also known as Newly Industrialized Nations.

Leaky PBX The ability of users to interconnect on-premises equipment to the public switched telecommunications network, thereby linking private lines, designed to link two points only, with often unmetered switched local services.

LEC Local Exchange Carrier. The provider of local services, often considered a natural monopoly and bottleneck through which most interexchange traffic traverses.

LEOs Low Earth Orbiting satellites. Satellites operating in orbits below the geostationary orbiting arc where service can be provided to low powered handheld terminals. Big LEOs provide a variety of voice and data services while small LEO provide data and emergency position location services.

level competitive playing field The goal of market entrants who claim that regulators and other decision-makers must take affirmative steps to reduce unfair market access opportunities accruing to incumbent carriers as a function of their earlier market entry, bottleneck control, customer base, and regulatory status.

liberalization The relaxation of rules and service obligations. For example, subsidized and under-priced local services, imposed on incumbent carriers, often occurring contemporaneously with privatization, deregulation, and market entry initiatives.

line-side interconnection Interconnection that takes place at a low level in the switching hierarchy typically for end users.

link budget The determination whether an adequate signal can be received by calculating the transmission power and factors that attenuate the signal.

link margin The estimated degree of signal strength above noise at a particular location under various circumstances.

local area network A network of work stations and personal computers linked via wiring installed by the operator or provided by the telephone company, or via wireless applications.

local loop Local exchange facilities comprising the first and last legs of a telecommunications route.

low Earth orbit The use of nongeostationary orbits for satellites primarily providing mobile telecommunications to lightweight transceivers. LEO satellites operate only a few hundred miles above the Earth surface, thereby reducing launch costs and the power needed to transmit to and from the satellites. On the other hand, the closer proximity to Earth requires a larger number of satellites to achieve desired regional or global coverage.

LRIC Long Run Incremental Cost. A measure of the additional cost incurred over the long term to provide a particular amount of a good or service. This measure accounts only for the additional costs incurred without regard to embedded fixed costs that could be shared by additional users.

market power The ability to affect the price or supply of a good or service.

MAOU Minimum Assignable Ownership Unit. The smallest unit of submarine cable capacity, typically a 64-Kbps circuit with associated capacity for multiplexing, allocatable to an investor or available to others through an Indefeasible Right of Use conveyance.

MEO Middle Earth Orbit. Nongeostationary satellite orbits between low Earth orbits below and geostationary orbits above where operators attempt to balance the benefits accruing from operating closer to Earth (e.g., access by lower powered terminals yet operating high enough to provide broad geographical coverage).

MFJ Modification of Final Judgment. Modifications to the 1956 Consent Decree agreed to by AT&T and the Justice Department thereby settling an antitrust suit. The MFJ required AT&T to divest its Bell Operating Companies in exchange for the opportunity to serve data processing markets. The spun-off Bell Operating Companies were limited to local exchange services.

mid-ocean The concept that international half-circuits are matched by foreign correspondents at the centerpoint of a submarine cable.

MITI Ministry of Trade and Industry. The powerful Japanese Ministry with trade and development portfolios.

modem A device that modulates and demodulates data signals enabling computers and other digital devices to operate over an analog network.

most favored nation A basic trade principle requiring nations associating with the WTO to extend to all nations any market access benefit accorded other nations.

MOU Memorandum of Understanding. A preliminary agreement setting forth basic terms and conditions typically followed up with a more comprehensive agreement.

MPT Ministry of Posts and Telecommunications. The government agency in many nations with telecommunications operational and regulatory responsibilities.

MSS Mobile Satellite Service. An ITU-recognized service designation for the provision of voice and data services to mobile users via geostationary or nongeostationary orbiting satellites.

MTS Message Telephone Service. Conventional dial-up voice telephone service also known by the acronym POTS (Plain Old Telephone Service).

multimedia The integration of various media previously considered available via a separate pipeline or marketing channel (e.g., using the television for entertainment and new interactive data processing, consumer order entry, and utility monitoring).

multinational enterprise A venture doing business in a number of nations making it a candidate for turnkey services from systems integrators and outsourcers.

multiplexing Subdividing a circuit into more than one channel to derive more throughput and capacity.

multipoint Using telecommunications services to serve more than one physical location (e.g., originating a video program for delivery to numerous cable systems using the broad geographical coverage of a satellite).

NAFTA North American Free Trade Agreement. A trade agreement between the United States, Canada, and Mexico resolving to reduce or eliminate barriers to trade between nations with an eye toward creating a regional trading bloc.

national hero A domestic carrier or manufacturer benefiting from policies designed to protect it from foreign competition including closed procurements and other policies that handicap, or prohibit, market entry.

national treatment Subjecting foreign manufacturers and service providers to the same regulatory treatment as afforded domestic enterprises.

natural monopoly A single manufacturer or service provider who singularly can operate most efficiently and who can maximize scale economies.

NIC Newly Industrialized Country. A nation with an expanding economy and, typically, with growing telecommunications requirements.

NII National Information Initiative. A plan for using government to stimulate primarily private ventures that will upgrade and expand the telecommunications and information-processing infrastructure in the United States. The government initiative includes a vision for ubiquitous access to a feature-rich information highway.

NIST National Institute of Science and Technology. An agency of the U.S. Department of Commerce that coordinates standard setting for the federal government and whose standards affect the parallel private standard-setting process.

node A point in a telecommunications or information-processing network configuration where two or more lines, routes, or pathways come together (e.g., at a switch, Earth station, or PBX).

noncommon carrier A classification that reduces or eliminates regulation in recognition of the carrier's lack of market dominance and the nonessentialness of the services offered.

NTIA National Telecommunications and Information Administration. The U.S. executive agency, located in the Department of Commerce, that serves as the President's researcher and advisor on telecommunication and information processing policy issues.

NTB Non-Tariff Barrier. A barrier to trade that is not documented in a customs duty or tariff.

NTSC National Television Standards Committee. The developer of a broadcast color television standard in the United States and implemented in North America and Japan.

NTT Nippon Telephone and Telegraph. The dominant local exchange and long distance carrier in Japan.

number portability The ability to designate a single telephone number for accessing individuals with several telecommunications devices operating in different locales.

NWIO New World Information Order. A vision for establishing closer parity of access to information between developed and developing nations.

OECD Organization for Economic Cooperation and Development. An organization that supports shared views on economic development, including compiling statistics and polling member nations on telecommunications and information policies.

offshore branching unit A device, located on the ocean floor, that is used to interconnect submarine cables, thereby expanding coverage and providing for redundant or alternative routing of traffic.

Oftel Office of Telecommunications. The United Kingdom independent regulatory agency.

one+ dialing A calling arrangement for direct connections with a prearranged billing commitment.

one-stop shopping The provision of a number of functions, like telecommunications network, design, procurement, and management by a single enterprise as an alternative to the end user performing these functions or securing the services from a number of enterprises.

OPEC Organization of Petroleum Exporting Countries. An organization seeking to manage the price and supply of petroleum.

open network architecture The FCC-articulated blueprint by which the Bell Operating Companies will revamp their facilities to provide for equal access to network facilities by affiliated and unaffiliated enhanced service providers.

open network provision The EC-articulated blueprint for harmonizing local exchange facilities access within the Community.

orbital arc Locations above Earth where satellites are located after launch. Most communication satellites occupy the geostationary orbital arc to maximize stability and geographical coverage.

OSI Open Systems Interconnection. A seven-level model for organizing data processing and telecommunications functions.

out-of-band signaling The use of a channel, separate from the one carrying traffic, to set up, supervise, and bill the call.

packet switching A transmission technology that reduces messages and data into individually routed packets and reassembles them before reaching the final destination.

PAL Phased Alternation by Line. A broadcast color television standard implemented in Europe and Asia.

PanAmSat Pan American Satellite. The first licensed U.S. facilities-based international satellite carrier that competes with INTELSAT. In 1997, Hughes Communications acquired the company.

party A nation that has acceded to an international agreement, thereby accepting its terms.

PBX Private Branch Exchange. A telephone line switching device located on the premises of users with a number of telephones and lines. The PBX switches and routes all inbound and outbound calls.

PCS Personal Communication Services. A variety of new terrestrial and satellite-delivered services to handheld terminals operating at low power.

peering A commercial arrangement among Internet service providers to accept traffic from onward routing at no cost.

photonics The use of specially treated glass fibers, lasers, and mirrors to provide a medium for extremely high-speed, high-capacity telecommunications and information processing.

point-to-multipoint Traffic routing from a single location to many recipients (e.g., video programming distribution to numerous cable operators).

portal A point of access to the Internet and specific content sources. Portals like Yahoo! and America Online simplify the process of accessing content by operating as a one-stop shop.

POTS Plain Old Telephone Service. Conventional telephone services provided to residential and small business users.

power flux density A measure of power radiated from a transmitter, like a satellite.

preemption Assertion of jurisdiction by one regulatory authority that would replace and dislodge the assertion of jurisdiction by another agency.

price cap A form of incentive regulation where carriers and regulators agree to replace conventional rate base regulation for a ceiling on customer charges and the requirement that these rates decline by a certain percentage each year, to reflect productivity improvements. The required annual reductions are possibly offset by a measure of overall increased costs to producers. Carriers are permitted to capture all or part of increased profits.

primary rate interface The large standard unit of capacity in an Integrated Services Digital Network comprised of twenty-three 64-Kbps bearer channels and one 16-Kbps data channel totaling 1.54 Mbps.

primary status The designation for a particular radio service that qualifies users for maximum permissible protection from interference by users of services holding a lower priority.

private line Dedicated capacity designed to link a single user with requirements in two international locations. Carriers now provide "virtual" private lines by partitioning capacity from public networks through the application of software.

privatization Wholly or partially converting the incumbent PTT to private ownership. The privatized enterprise may maintain a monopoly or face varying degrees of competition from market entrants.

protocol A standard operating procedure or format.

PSTN Public Switched Telecommunications Network. The publicly available local and long haul facilities of the incumbent carrier.

PTO Public Telecommunication Organization. A new designation for Post Telegraph and Telephone administrations.

PTT Post Telegraph and Telephone administration. The incumbent and often exclusive carrier in most nations that provides services on a ubiquitous basis, some of which may be subsidized.

RACE Research and Development in Advanced Communications in Europe. An initiative in the European Community to expedite and emphasize telecommunications research and development.

Ramsey pricing Charging users on the basis of their demand elasticity. For example, users with plenty of options, including dedicated facilities and leased lines from a number of carriers, qualify for rates below those charged to users with fewer options.

RASCOM Regional African Satellite Communication System. An affiliation of African telecommunications administrations with plans to operate a dedicated satellite system after having leased capacity from INTELSAT.

RBOC Regional Bell Operating Company. Upon divestiture from AT&T, the Bell Operating Companies were reformulated into seven new corporations, each representing a particular geographical region of the United States.

redundancy The availability of backup capacity to restore service in the event of a facility outage or peak demand condition.

resale The acquisition of bulk transmission capacity and other services for subsequent resale to individual users who singularly do not generate the demand for large capacity offerings. Resellers perform an arbitrage function and profit by acquiring discounted bulk capacity from underlying facilities- based carriers and repackaging it at rates less than what an individual user could secure directly from the carrier.

RFP Request For Proposals. A tender offer soliciting bids by parties interested in securing a contract to perform some form of work.

roamer An individual desiring to use a mobile telecommunications device, perhaps a cellular radiotelephone, while away from the location where service is usually provided.

rolloff The manner in which signal strength drops off at locations increasingly distant from the targeted service location.

RPOA Recognized Private Operating Agency. An ITU-approved designation of nongovernmental telecommunications service providers that confers official recognition and enhances the ability of such organizations to participate in ITU forums.

rulemaking The process by which a U.S. administrative or regulatory agency establishes a binding rule, regulation, or policy. Procedural due process requirements obligate the agency to notify the public of proposed actions and to provide opportunities for participation through filed comments or testimony.

satellite beam types There are four satellite beam types: Global, Hemispheric, Zone, and Spot. Global: Transmission beams with maximum geographical illumination, approximately one-third of the entire globe. Hemispheric: Transmission beams shaped to cover an entire hemisphere (e.g., North America, Central and South America, or Europe). Zone: Transmission beams shaped to cover one or more portions of a hemisphere with higher signal strength. Spot: Transmission beams shaped to concentrate signal strength over relatively small, but highly populated locations.

S-band The portion of the frequency spectrum around 2 GHz used for terrestrial microwave and some mobile satellite applications.

SDR Special Drawing Rights. An aggregation of national currencies administered by the International Monetary Fund and often used by international carriers in settling toll revenue accounts.

SECAM Sequential Color with Memory. A French-developed broadcast television standard adopted in France, French-speaking African nations, and the former Soviet Union.

sender keep all A toll revenue division arrangement whereby the originating carrier keeps all charges.

sender pay all A charging mechanism whereby the calling party pays all charges.

separate system A satellite system, separate from the INTELSAT cooperative, providing international services.

server A device that houses content accessed via the Internet.

settlement rate The unit of currency and percentage division negotiated by carriers for settling accounts regarding carriage of inbound international traffic. Typically, international carriers on direct routings equally divide the negotiated rate.

signatory A company or government that has executed a binding legal commitment (e.g., Comsat serves as the sole U.S. signatory to INTELSAT and Inmarsat).

Single Europe Act Legislation adopted by the constituent members of the European Community agreeing to closer collaboration and harmonization of laws, policies, and regulations.

smart card A credit-card–sized instrument containing microchips and associated electronics for providing credit and other data for facilitating transactions.

software defined network The use of software to partition transmission capacity from public networks for private use, making it appear as the functional equivalent to capacity dedicated for a single user.

SONET Synchronous Optical Network. Protocols for operating fiber optic transmission facilities at very high, multimegabit speeds.

sovereignty Recognition of independence and legitimacy in a nation's determination of policies, laws, and regulations affecting its citizens and internal affairs.

space station A satellite or other device that operates in space.

spectrum A term depicting the range of frequencies that are used in telecommunications.

spectrum allocation The designation of a frequency band to one or more specified services by the International Telecommunication Union and by individual nations.

spectrum allotment The designation of specific channels of operation to particular localities.

spectrum assignment The grant of operational authority that includes a license to operate on a specific frequency.

spread spectrum A transmission technique whereby signals are transmitted over a range of spectrum at low power with messages handled by assigning a discrete sequence of code or by hopping over different frequencies.

SS7 Signaling System 7. An advanced method for providing software defined signaling necessary to set up telecommunications links.

station keeping The use of jets aboard a satellite to keep it "on station" (in the prescribed orbital location with antennas properly aimed). A station-kept satellite eliminates costly tracking requirements for each Earth station pointing to the satellite. Satellites typically reach end of life when station-keeping fuel has been used up, even though the electronic and power generation components remain operational. Operators may conserve station-keeping fuel to extend a satellite's usable life through the use of maneuvers that allow the satellite to drift slightly.

STM-1 Synchronous Transport Module Ones. A unit of submarine cable transmission capacity equivalent to 155 Mbps of throughput.

store-and-forward The use of computerization to store traffic temporarily for subsequent transmission in batches or efficiently loaded packets. A satellite may store messages until it reaches a point in its orbit where the traffic can be downlinked.

strategic alliances Corporate business relationships designed to tap the comparative advantages and expertise of the participants as well as to achieve a broader geographical marketing presence.

Submarine Cable Landing License Act An act of Congress requiring the FCC and State Department to grant licenses to land a submarine cable on U.S. soil. Foreign carriers will receive such grants only if U.S.-owned and operated submarine cables can make landfall in the foreign country.

substitutable products Products that can replace and compete with each other (e.g., tea instead of coffee and Centrex instead of private branch exchanges).

switch The portion of the telecommunications infrastructure where traffic is received by a device that identifies the intended destination and selects the routing for delivering traffic onward to that point.

T-1 A unit of transmission capacity equal to 1.544 Mbps in throughput.

tail circuits Domestic facilities used to transport traffic from an international caller to a gateway, and from another gateway to the call recipient.

tandem switch A telecommunications switching facility with management control over one or more end offices.

tariff A contract for service setting out the terms and conditions under which the general public will secure service.

TCP/IP Transmission Control Protocol/Internet Protocol. The basic technical standards for Internet traffic routing and addressing.

TDMA Time Division Multiple Access. A technique for deriving more throughput and accommodating increased demand by assigning channels to users in discrete blocks of time.

Telecommunications Act of 1996 An Act of the United States Congress that substantially revised and amended the Communications Act of 1934 to promote competition.

Telecommunications Development Bureau (BDT) A permanent organ of the International Telecommunication Union formed in 1989 to promote telecommunications development, including technical and managerial training.

teledensity The penetration of telecommunications access lines typically measured per 100 inhabitants.

telepoint A new generation of pay telephones and one-way wireless communications based on microcellular radio technology.

teleport A satellite Earth station with extensive access to terrestrial facilities and a broad base of users, situated on real estate developments adjacent to the Earth stations or at other locations.

teletext Slow speed transmission of textual information over the vertical blanking intervals of broadcast television signals.

throughput The amount of data carried over a particular amount of capacity in a specified time span (see also bit rate).

traffic The messages, programming, files, and other intelligence requiring transport from one location to another.

transborder data flow The transmission of data across national borders, alternatively viewed as a vehicle to share the wealth of information, or a source for cultural imperialism.

transceiver A device that contains components that transmit and receive radio signals.

transiting The carriage of international traffic via the facilities of an intermediary, third country between the sender and recipient.

transponder The components in a satellite able to transmit and receive traffic. Satellite capacity is typically stated in the number of available 36 MHz equivalent transponders.

Treaty of Rome The treaty enacted in 1957 that created the European Community.

trunking A computer assisted process for queuing channel demands thereby more efficiently using spectrum and accommodating episodic demands for service. For example, a trunked radio system will provide channels of communications for users on an as-demanded basis from the available inventory.

trunk side interconnection The interconnection of facilities and lines between carriers at a level higher up the hierarchy than typically accorded end users.

TVRO Television Receive-Only. Satellite dishes equipped for one-way reception of television signals.

unbundled Separation of services and equipment into discrete elements available on an individual, à la carte basis.

uplink The process of transmitting traffic from Earth to a satellite.

USISC United States International Service Carrier. A U.S. carrier licensed by the FCC to provide international services.

USTR United States Trade Representative. The U.S. Executive Branch agency responsible for engaging foreign nations in bilateral and multilateral trade negotiations.

VAN Value Added Network. An enhanced services provider that typically leases transmission capacity and adds value by performing customized services (e.g., credit card verification and electronic funds transfer).

video dial tone The provision of transparent common carrier transmission capacity for delivery of video programming.

videotext Slow speed transmission of textual information over closed circuits.

virtual networks Software derived networks typically involving lines partitioned from public switched facilities.

virtual private line Using software to partition capacity from switched public facilities to make it function the same as capacity dedicated to a single user.

VSAT Very Small Aperture Terminal. Very small satellite Earth stations, typically situated on user premises, that promote diversification in services, easier access, and lower costs.

WAN Wide Area Network. An interconnected network of personal computers and work stations situated in different locations.

WARC/WRC World Administrative Radio Conference/World Radio Conference. Periodically convened conferences of ITU member nations to address regional or global spectrum planning and policy issues.

WATS Wide Area Telephone Service. Toll-free calling to the call originator provided by most retailing and service organizations.

whipsawing The practice of a monopoly carrier to extract concessions from foreign correspondent carriers operating in a competitive market who must vie for return inbound traffic from the monopoly carrier.

windowing Segmenting the distribution of a product (e.g., a movie) into sequences based on consumer demand elasticities and time from initial availability. Movie producers attempt to extract the greatest amount of rent by calibrating availability and charges: $6–8 at a theater, $2–3 on rented video tape, and no charge to viewers on broadcast television.

wireless local loop The use of unwired transmission facilities to provide connections to the wireline public-switched telecommunications network; many of these technologies also provide mobile services.

wireless networks The use of radio technology to serve increasingly diverse applications, including mobile services and some applications heretofore provided via wireline facilities.

wireline The conventional wire-based telecommunications infrastructure installed and managed by incumbent carriers.

World Wide Web That portion of the Internet that combines text, sound, video, and other graphical material accessible via broswers.

WTO World Trade Organization. The permanent trade policy and dispute resolution forum created in 1994 by nations that had formed the GATT.

WTTC World Telephone and Telegraph Conference. A periodically convened conference of ITU member nations to address issues pertaining to the rules of the road in telephony and other telecommunications services.

X.25 An ITU-recognized protocol for user access to data networks.

X.75 An ITU-recognized protocol for interconnecting data networks.

X.400/X.500 A series of protocols governing electronic mail and directory functions.

X-band A portion of the frequency spectrum in the 7–8 GHz range used primarily for satellites with defense and intelligence gathering applications.

zero+ dialing A calling arrangement involving verification of credit or payment through intervention by an operator or computer.

About the author

Jennifer A. Manner is currently the director of international alliances at WorldCom, Inc., in Washington, D.C., where she is responsible for crafting and implementing market access strategies for a wide variety of products for the company. Prior to this position, Ms. Manner was the associate counsel and focus group leader for Foreign Market Access and International Wireless Services for WorldCom. Before joining WorldCom, Ms. Manner was an associate at the law firm of Akin, Gump, Strauss, Hauer, and Feld in Washington, D.C., and an attorney-advisor at the Federal Communications Commission.

Ms. Manner is currently an adjunct professor of law at Georgetown University Law Center, where she teaches international telecommunications regulation. Ms. Manner is also active in many professional organizations and has held key leadership positions in these groups, including serving on the board of directors of the U.S. International Telecommunications Union Association; chairing the Washington, D.C., Bar Association's Computer and Telecommunications Law Section; serving on the board of directors of the Federal Communications Bar Association; chairing several committees, including the International Committee; and being a member of the International Space Law Society.

Ms. Manner received a master of law in international law with distinction from Georgetown University Law Center, a juris doctorate *cum laude*, from New York Law School, and a B.A. in theater arts and political science from the State University of New York at Albany. She lives in Kensington, Maryland, and Kent, Ohio, with her husband, Dr. Eric Glasgow, a professor of neurobiology at the Northeastern Universities of Ohio College of Medicine.

Index

C

D

E

F

G

H

I

J

L

M

W

Artech House Communications Law and Policy Series

Global Telecommunications Market Access, Jennifer A. Manner

International Telecommunications Handbook, Robert M. Frieden

An Introduction to U.S. Telecommunications Law, Second Edition, Charles H. Kennedy

The Law and Regulation of Telecommunications Carriers, Henk J. Brands and Evan T. Leo

A Manager's Guide to the Law and Economics of Data Networks, Jeffrey H. Matsuura

Managing Internet-Driven Change in International Telecommunications, Rob Frieden

Practical Internet Law for Business, Kurt M. Saunders

Securities, Rights, and Liabilities in E-Commerce, Jeffrey H. Matsuura

Telecommunications Deregulation and the Information Ecomomy, Second Edition, James K. Shaw

For further information on these and other Artech House titles, including previously considered out-of-print books now available through our In-Print-Forever® (IPF®) program, contact:

Artech House
685 Canton Street
Norwood, MA 02062
Phone: 781-769-9750
Fax: 781-769-6334
e-mail: artech@artechhouse.com

Artech House
46 Gillingham Street
London SW1V 1AH UK
Phone: +44 (0)171-973-8077
Fax: +44 (0)171-630-0166
e-mail: artech-uk@artechhouse.com

Find us on the World Wide Web at:
www.artechhouse.com